THE A & P
COLORING WORKBOOK

Essentials of Human Anatomy & Physiology

THIRD EDITION

Now in full color, this successful text uses a clear and friendly writing style to distill important material about the structure and function of the human body. The updated art program of over 200 pieces is clear and colorful, attractive and accurate. Other features of the Third Edition include:

- Body system sections color-tabbed for quick reference
- Balanced anatomy, physiology, and clinical coverage
- Interesting analogies to explain difficult concepts
- New chemistry chapter
- Exceptionally strong immunology chapter
- Special Topic Boxes to show integration of structure and function
- An abundance of clinical examples, with emphasis on homeostasis
- Ample learning aids for each chapter

The Perfect Companion to Your Coloring Workbook —Also by Elaine N. Marieb

Contents

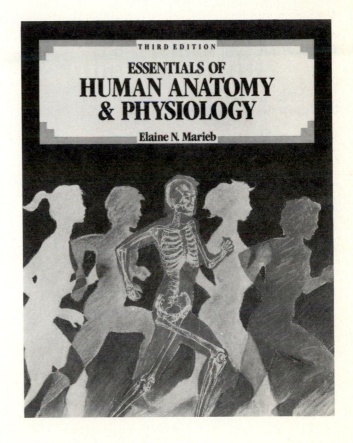

Softcover, 8½″ × 11″, 450 pages, ISBN 0-8053-4804-2.
Available at your college or university bookstore. Or, to order directly from
Benjamin/Cummings, call our Order Department at (800) 447-2226.

The Benjamin/Cummings Series in Human Anatomy and Physiology

R. A. Chase
The Bassett Atlas of Human Anatomy (1989)

S. W. Langjahr and R. A. Brister
Human Anatomy: A Computerized Review and Coloring Atlas (1985)

E. N. Marieb
Human Anatomy and Physiology (1989)

E. N. Marieb
Human Anatomy and Physiology Lab Manual: Cat Version, Third Edition (1989)

E. N. Marieb
Human Anatomy and Physiology Lab Manual: Fetal Pig Version, Third Edition (1989)

E. N. Marieb
Human Anatomy and Physiology, Lab Manual: Brief Version, Second Edition (1987)

E. N. Marieb
Essentials of Human Anatomy & Physiology, Third Edition (1991)

E. N. Marieb
The A & P Coloring Workbook: A Complete Study Guide, Third Edition (1991)

E. B. Mason
Human Physiology (1983)

A. P. Spence
Basic Human Anatomy, Third Edition (1991)

R. L. Vines and A. Hinderstein
Human Musculature Videotape (1989)

THE A & P COLORING WORKBOOK

A Complete Study Guide

THIRD EDITION

Elaine N. Marieb, R.N., Ph.D.

Holyoke Community College

The Benjamin/Cummings Publishing Company, Inc.

Redwood City, California • Fort Collins, Colorado • Menlo Park, California
Reading, Massachusetts • New York • Don Mills, Ontario • Wokingham, U.K.
Amsterdam • Bonn • Sydney • Singapore • Tokyo • Madrid • San Juan

Sponsoring Editor: Melinda Adams

Production Supervisor: Brian Jones

Marketing Manager: Stacy Treco

Text Designer: John Edeen

Cover Designer: Gary Head

Cover Photographer: David Wakely

Illustrators: Sibyl Graber-Gerig;
Reese Thornton, Folium;
Pauline Phung

Copyeditor: Jenny Hale Pulsipher

Proofreaders: Rosana Francescato,
Melissa Andrews

Compositor: Fog Press

ISBN 0-8053-4806-9

6 7 8 9 10 -AL- 95 94 93

The Benjamin/Cummings Publishing Company, Inc.
390 Bridge Parkway
Redwood City, California 94065

Preface

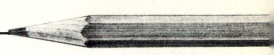

Although never a simple task, the study of the human body is always fascinating. The Third Edition of *The A & P Coloring Workbook* continues to serve as a review and reinforcement tool to help health professional and life-science students master the basic concepts of human anatomy and physiology.

Scope

Although this book reviews the human body from microscopic to macroscopic levels (that is, topics range from simple chemistry and cells to body organ systems), it is not intended to be encyclopedic. In fact, to facilitate learning, this *Workbook* covers only the most important and useful aspects of human anatomy and physiology. Pathophysiology is briefly introduced with each system so that students can apply their learning. Where relevant, clinical aspects (for example—muscles used for injection sites, the role of ciliated cells in protection of the respiratory tract, and reasons for skin-ulcer formation) are discussed. To encourage conceptualization of the human body as a dynamic and continually changing organism, developmental aspects of youth, adulthood, and old age are included. There's also a separate chapter on the immune system, covering topics of increasing importance to future health professionals.

Learning Aids

As in the first and second editions, multiple pedagogical devices are used throughout the book to test comprehension of key concepts. The integration of a traditional study-guide approach with visualization and coloring exercises is unique. A variety of exercises demands learning on several levels, avoids rote memorization, and helps maintain a high level of interest.

The exercises include *completion* from a selection of key choices, *matching* terms or descriptions, and *labeling* diagrams. *Elimination* questions require the student to discover the similarities or dissimilarities among a number of structures or objects and to select the one that is not appropriate. *Correctible true/false* questions add a new dimension to the more traditional form of this exercise. Also, students are asked to provide important *definitions*. In the completion sections, the design of this edition provides longer answer lines so that the student can write in either the key letter *or the appropriate term*. Both responses are then given in the answer section.

Coloring exercises have been shown to be a motivating, effective approach to learning. Each illustration has been carefully prepared to show sufficient detail for learning without students becoming bored with coloring. There are 100 coloring exercises throughout the text that should prove valuable to all students. Those students who are visually oriented will find these exercises particularly beneficial. When completed, the color diagrams provide an ideal reference and review.

Visualization exercises are another unique feature of this book. With the exception of the introductory chapter on terminology, each chapter ends with an "Incredible Journey." Students are asked to imagine themselves in miniature, traveling within the body through various organs and systems. These visualization exercises are optional, but they often summarize chapter content, allowing students to assimilate what they have learned in unusual and amusing ways.

Acknowledgments

To those educators, colleagues, and students who have provided feedback and suggestions during the preparation of all three editions of this workbook, I am sincerely grateful. I also wish to acknowledge with gratitude the excellent art work provided by Sibyl Graber-Gerig, Reese Thornton of Folium, and Pauline Phung; there is little doubt that it has enhanced this work. The staff at Benjamin/Cummings has continuously supported my efforts to turn out a study tool that will be well received and beneficial to both educator and student audiences. I particularly wish to thank Brian Jones for his work as production supervisor on this edition. His editorial wisdom and ability to see (and circumvent) possible problems before they happened, certainly made my life more pleasant during the production process. I also owe a debt of gratitude to my editors, Connie Spatz and Melinda Adams, for their diligence on this project.

Finally, and as ever, I want to thank my husband, Joseph, for his unwavering support.

Instructions for the Student —How to Use This Book

Dear Student,

The A & P Coloring Workbook has been created particularly for you. It is the outcome of years of personal attempts to find and create exercises most helpful to my own students when they study and review for a lecture test or laboratory quiz.

Although I never cease to be amazed at how remarkable the human body is, I would never try to convince you that studying it is easy. The study of human anatomy and physiology has its own special terminology; it requires that you become familiar with the basic concepts of chemistry to understand physiology; and often (sadly) it requires rote memorization of facts. It is my hope that this workbook will help simplify your task. To make the most of the exercises, read these instructions carefully before starting work.

Labeling and coloring. Some of these questions ask you only to label a diagram, but most also ask that you do some coloring of the figure. You can usually choose whichever colors you prefer. Soft, colored pencils are recommended so that the underlying diagram shows through. Since most figures have multiple parts to color, you will need a variety of colors—18 should be sufficient. In the coloring exercises, you are asked to choose a particular color for each structure to be colored. That color is then used to fill in both a color-coding circle found next to the name of the structure or organ, and the structure or organ on the figure. This allows you to identify the colored structure quickly and by name in cases where the diagram is not labeled. In a few cases, you are given specific coloring instructions to follow.

Matching. Here you are asked to match a term denoting a structure or physiologic process with a descriptive phrase or sentence. Depending on how you like to review the material, you can use either the terms from the key choices or their associated key letters. Both answers are provided in the answer section.

Completion. You select the correct term to answer a specific question, or you fill in blanks to complete a sentence. In many exercises, some terms are used more than once and others are not used at all.

Definitions. You are asked to provide a brief definition of a particular structure or process.

True or False. One word or phrase is underlined in a sentence. You decide if the sentence is true as it is written. If not, you correct the underlined word or phrase.

Elimination. Here you are asked to find the term that does not "belong" in a particular grouping of related terms. In this type of exercise, you must analyze how the various terms are similar to or different from the others.

Visualization. "The Incredible Journey" is a special type of completion exercise, which ends every chapter except the first one. For this exercise, you are asked to imagine that you have been miniaturized and injected into the body of a human being (your host). Anatomic landmarks and physiologic events are described from your miniaturized viewpoint, and you are then asked to identify your observations. Although this exercise is optional, my students have found them fun to complete, and I hope you will too.

Each exercise has complete instructions, which you should read through carefully before beginning the exercise. When there are multiple instructions, complete them in the order given.

At times, it may appear that information is being duplicated in the different types of exercises. Although there is some overlap, the understandings being tested are different in the different exercises. Remember, when you understand a concept from several different perspectives, you have mastered that concept.

I sincerely hope that *The A & P Coloring Workbook* challenges you to increase your knowledge, comprehension, retention, and appreciation of the structure and function of the human body.

Good luck!

Elaine Marieb

Contents

The Human Body—
An Orientation

Most of us have a natural curiosity about our bodies, and the study of anatomy and physiology elaborates on this interest. Anatomists have developed a universally acceptable set of reference terms that allows body structures to be located and identified with a high degree of clarity. Initially, students might have difficulties with the language used to describe anatomy and physiology, but without such a special vocabulary, confusion is inevitable.

The topics in this chapter enable students to test their mastery of terminology commonly used to describe the body and the relationships of its various parts. Concepts concerning functions vital for life and homeostasis are also reviewed. Additional topics include body organization from simple-to-complex levels and an introduction to the organ systems forming the body as a whole.

An Overview of Anatomy and Physiology

1. Match the terms in Column B to the appropriate descriptions provided in Column A. Enter the correct letter or its corresponding term in the answer blanks.

Column A

_____ **1.** The branch of biologic science that studies and describes how body parts work or function

_____ **2.** The study of the shape and structure of body parts

_____ **3.** The tendency of the body's systems to maintain a relatively constant or balanced internal environment

_____ **4.** The term that indicates *all* chemical reactions occurring in the body.

Column B

A. Anatomy

B. Homeostasis

C. Metabolism

D. Physiology

2. Circle all the terms or phrases that correctly relate to the study of *physiology,* use a highlighter to identify those terms or phrases that pertain to *anatomical studies.*

   A. Measuring an organ's size, shape, and weight   **H.** Dynamic

   B. Can be studied in dead specimens   **I.** Dissection

   C. Often studied in living subjects   **J.** Experimentation

   D. Chemistry principles   **K.** Observation

   E. Measuring the acid content of the stomach   **L.** Directional terms

   F. Principles of physics   **M.** Static

   G. Observing a heart in action

Levels of Structural Organization

3. The structures of the body are organized into successively larger and more complex structures. Fill in the answer blanks with the correct terms for these increasingly larger structures.

Chemicals ⟶ _____ ⟶ _____ ⟶

_____ ⟶ _____ ⟶ Organism

4. Circle the term that does not belong in each of the following groupings.

   1. Atom   Cell   Tissue   Alive   Organ

   2. Brain   Stomach   Heart   Liver   Epithelium

   3. Epithelium   Heart   Muscle tissue   Nervous tissue
      Connective tissue

   4. Human   Digestive system   Horse   Pine tree   Amoeba

5. Using the key choices, identify the organ systems to which the following organs or functions belong. Insert the correct letter or term in the answer blanks.

   KEY CHOICES:

   A. Circulatory   **D.** Integumentary   **G.** Reproductive   **I.** Skeletal

   B. Digestive   **E.** Muscular   **H.** Respiratory   **J.** Urinary

   C. Endocrine   **F.** Nervous

_____ **1.** Rids the body of nitrogen-containing wastes

_____ **2.** Is affected by the removal of the thyroid gland

_____ **3.** Provides support and levers on which the muscular system can act

_____ **4.** Includes the heart

_____ **5.** Protects underlying organs from drying out and mechanical damage

_____ **6.** Protects the body; destroys bacteria and tumor cells

_____ **7.** Breaks down foodstuffs into small particles that can be absorbed

_____ **8.** Removes carbon dioxide from the blood

_____ **9.** Delivers oxygen and nutrients to the body tissues

_____ **10.** Moves the limbs; allows facial expression

_____ **11.** Conserves body water or eliminates excesses

_____ **12.** Allows conception and childbearing

_____ **13.** Controls the body with chemicals called hormones

_____ **14.** Is damaged when you cut your finger or get a severe sunburn

6. Using key choices from Exercise 5, choose the organ system to which each of the following sets of organs belongs. Enter the correct letter in the answer blanks.

_____ **1.** Blood vessels, spleen, lymphatic tissue

_____ **2.** Pancreas, pituitary, adrenal glands

_____ **3.** Kidneys, bladder, ureters

_____ **4.** Testis, vas deferens, urethra

_____ **5.** Esophagus, large intestine, rectum

_____ **6.** Breastbone, vertebral column, skull

_____ **7.** Brain, nerves, sensory receptors

7. Figures 1-1 to 1-6 represent the various body organ systems. First identify and name each organ system by filling in the blank directly under the illustration. Then select a different color for each organ and use it to color the coding circles and corresponding structures in the illustrations.

○ Blood vessels   ○ Nasal cavity

○ Heart   ○ Lungs

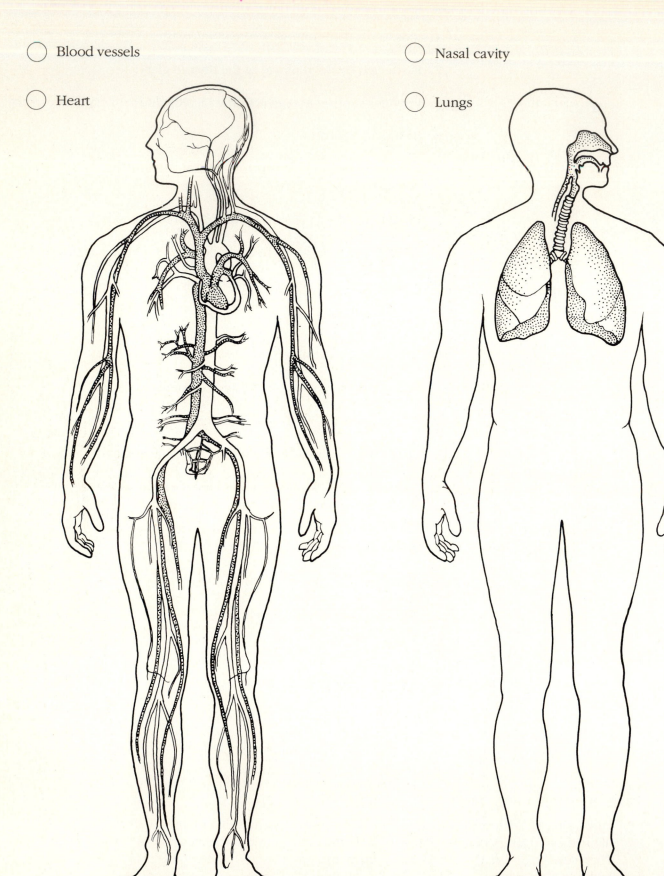

Figure 1-1   **Figure 1-2**

◯ Brain

◯ Spinal cord

◯ Nerves

◯ Kidneys

◯ Ureters

◯ Bladder

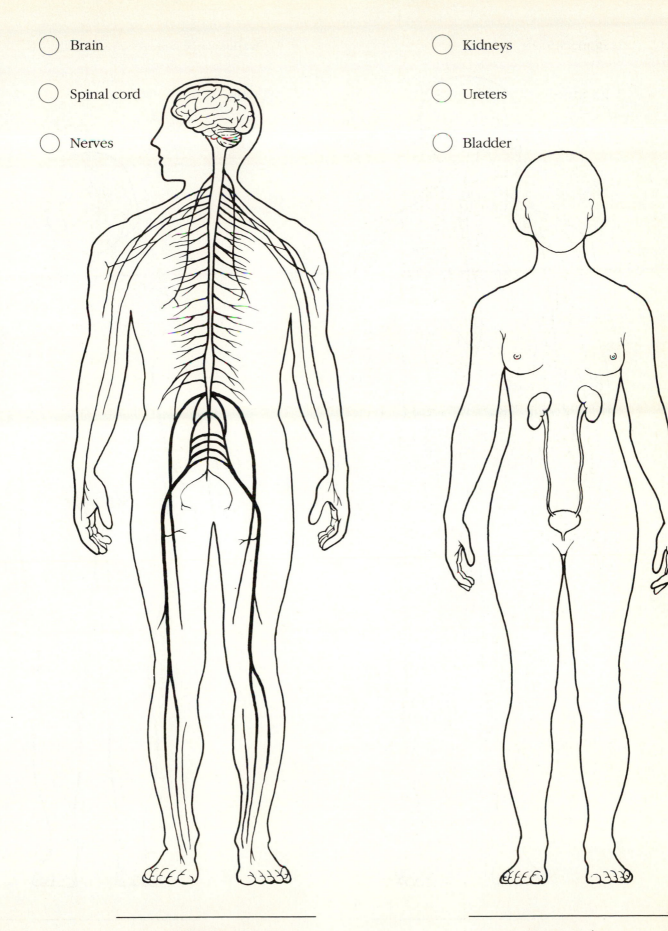

Figure 1-3

Figure 1-4

○ Stomach

○ Intestines

○ Ovaries

○ Uterus

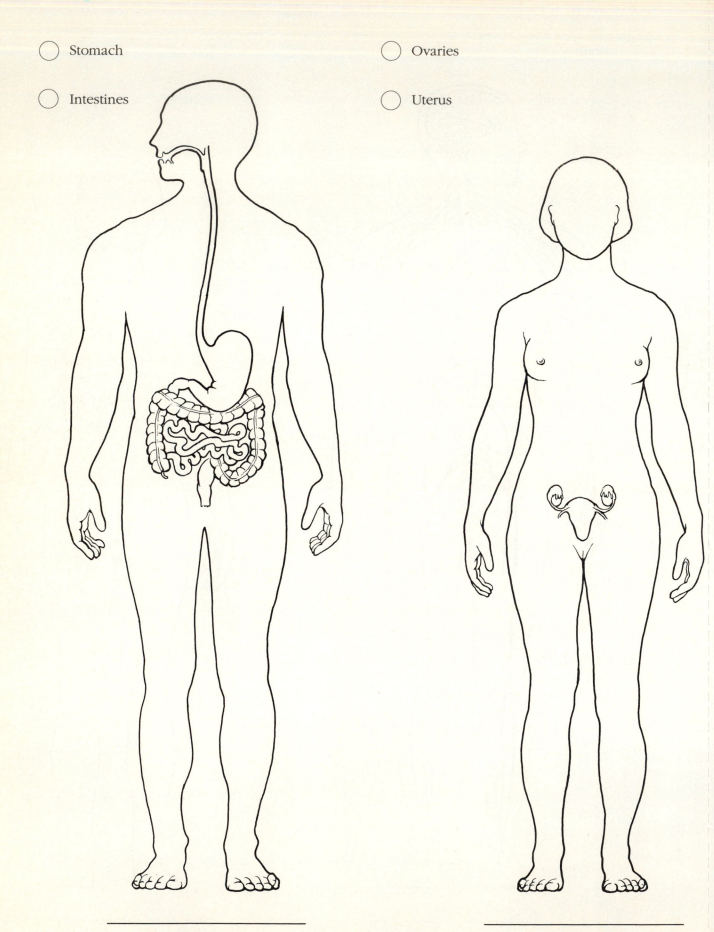

Figure 1-5

Figure 1-6

Maintaining Life

8. Match the terms pertaining to functional characteristics of organisms in Column B with the appropriate descriptions in Column A. Fill in the answer blanks with the appropriate letter or term.

Column A   Column B

_____   1. Keeps the body's internal environment distinct from the external environment

A. Digestion

B. Excretion

_____   2. Provides new cells for growth and repair

C. Growth

_____   3. Occurs when constructive activities occur at a faster rate than destructive activities

D. Maintenance of boundaries

_____   4. The tuna sandwich you have just eaten is broken down to its chemical building blocks

E. Metabolism

_____   5. Elimination of carbon dioxide by the lungs and elimination of nitrogenous wastes by the kidneys

F. Movement

G. Responsiveness

_____   6. Ability to react to stimuli; a major role of the nervous system

H. Reproduction

_____   7. Walking, throwing a ball, riding a bicycle

_____   8. All chemical reactions occurring in the body

_____   9. At the cellular level, membranes; for the whole organism, the skin

9. Using the key choices, correctly identify the survival needs that correspond to the following descriptions. Insert the correct letter or term in the answer blanks.

KEY CHOICES

A. Appropriate body temperature   **C.** Nutrients   **E.** Water

B. Atmospheric pressure   **D.** Oxygen

_____   1. Includes carbohydrates, proteins, fats, and minerals

_____   2. Essential for normal operation of the respiratory system and breathing

_____   3. Single substance accounting for over 60% of body weight

_____   4. Required for the release of energy from foodstuffs

_____   5. Provides the basis for body fluids of all types

_____   6. When too high or too low, physiological activities cease, primarily because molecules are destroyed or become nonfunctional.

Homeostasis

10. The following statements refer to homeostatic control systems. Complete each statement by inserting your answers in the answer blanks.

_____ 1.

_____ 2.

_____ 3.

_____ 4.

_____ 5.

_____ 6.

_____ 7.

_____ 8.

_____ 9.

There are three essential components of all homeostatic control mechanisms: control center, receptor, and effector. The __(1)__ senses changes in the environment and responds by sending information (input) to the __(2)__ along the __(3)__ pathway. The __(4)__ analyzes the input, determines the appropriate response, and activates the __(5)__ by sending information along the __(6)__ pathway. When the response causes the initial stimulus to decline, the homeostatic mechanism is referred to as a __(7)__ feedback mechanism. When the response enhances the initial stimulus, the mechanism is called a __(8)__ feedback mechanism. __(9)__ feedback mechanisms are much more common in the body.

The Language of Anatomy

11. Complete the following statements by filling in the answer blanks with the correct term.

_____ 1.

_____ 2.

_____ 3.

The abdominopelvic and thoracic cavities are subdivisions of the __(1)__ body cavity; the cranial and spinal cavities are parts of the __(2)__ body cavity. The __(3)__ body cavity is totally surrounded by bone and provides very good protection to the structures it contains.

12. Circle the term or phrase that does not belong in each of the following groupings.

1. Transverse   Distal   Frontal   Sagittal

2. Lumbar   Thoracic   Cubital   Abdominal   Scapular

3. Calf   Brachial   Femoral   Popliteal

4. Epigastric   Hypogastric   Right iliac   Left upper quadrant

13. Select different colors for the *dorsal* and *ventral* body cavities. Color the coding circles below and the corresponding cavities in part A of Figure 1-7. Complete the figure by labeling those body cavity subdivisions that have a leader line. Complete part B by labeling each of the abdominal regions indicated by a leader line.

◯ Dorsal body cavity   ◯ Ventral body cavity

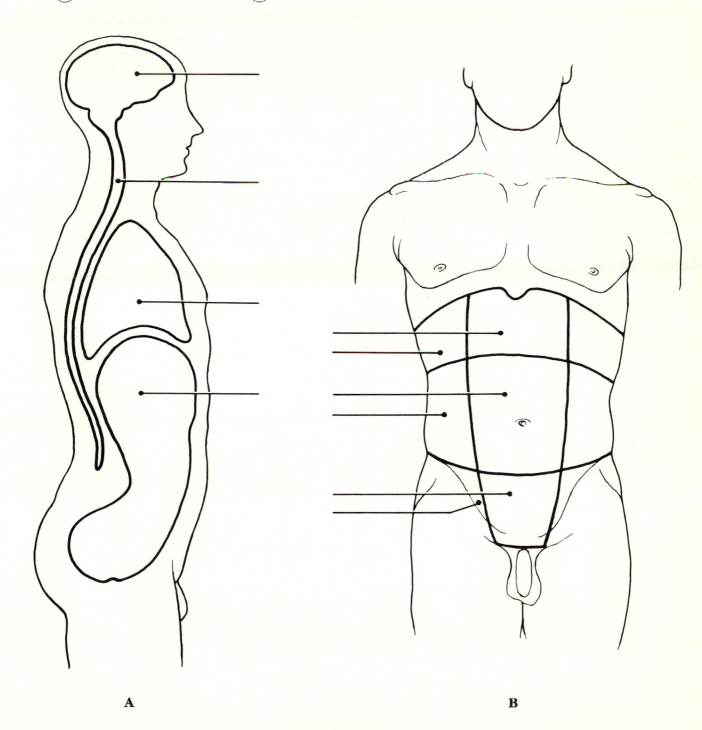

A   B

Figure 1-7

14. Select the key choices that identify the following body parts or areas. Enter the appropriate letter or corresponding term in the answer blanks.

    KEY CHOICES:

    A. Abdominal   **E.** Buccal   **I.** Gluteal   **M.** Popliteal

    B. Antecubital   **F.** Calf   **J.** Inguinal   **N.** Pubic

    C. Axillary   **G.** Cervical   **K.** Lumbar   **O.** Umbilical

    D. Brachial   **H.** Femoral   **L.** Occipital   **P.** Scapular

    _____ **1.** Armpit

    _____ **2.** Thigh region

    _____ **3.** Buttock area

    _____ **4.** Neck region

    _____ **5.** "Belly button" area

    _____ **6.** Genital area

    _____ **7.** Anterior aspect of elbow

    _____ **8.** Posterior aspect of head

    _____ **9.** Area where trunk meets thigh

    _____ **10.** Back area from ribs to hips

    _____ **11.** Pertaining to the cheek

15. Using the key terms from Exercise 14, correctly label all body areas indicated with leader lines on Figure 1-8.

    In addition, identify the sections labeled **A** and **B** in the figure.

    Section **A:** _____

    Section **B:** _____

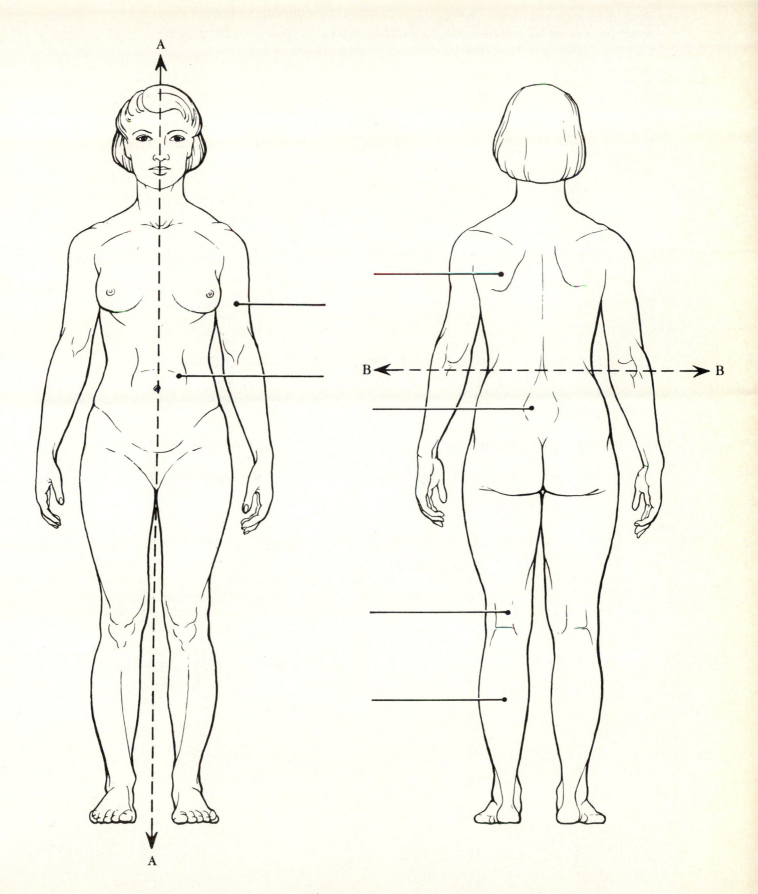

Figure 1-8

16. From the key choices, select the body cavities where the following surgical
     procedures would occur. Insert the correct letter or term in the answer blanks. Be
     precise: also select the name of the cavity subdivision if appropriate.

     KEY CHOICES:

     A. Abdominal   **C.** Dorsal   **E.** Spinal   **G.** Ventral

     B. Cranial   **D.** Pelvic   **F.** Thoracic

     _____ **1.** Removal of the uterus, or womb

     _____ **2.** Coronary bypass surgery (heart surgery)

     _____ **3.** Removal of a serious brain tumor

     _____ **4.** Removal of a "hot" appendix

     _____ **5.** A stomach ulcer operation

17. Complete the following statements by choosing an anatomical term from the
     key choices. Enter the appropriate letter or term in the answer blanks.

     KEY CHOICES

     A. Anterior   **D.** Inferior   **G.** Posterior   **J.** Superior

     B. Distal   **E.** Lateral   **H.** Proximal   **K.** Transverse

     C. Frontal   **F.** Medial   **I.** Sagittal

     _____ **1.**   In the anatomical position, the face and palms are on the
                                           __(1)__ body surface, the buttocks and shoulder blades are
     _____ **2.**   on the __(2)__ body surface, and the top of the head is the
                                           most __(3)__ part of the body. The ears are __(4)__ to the
     _____ **3.**   shoulders and __(5)__ to the nose. The heart is __(6)__ to the
                                           spine and __(7)__ to the lungs. The elbow is __(8)__ to the
     _____ **4.**   fingers but __(9)__ to the shoulder. In humans, the dorsal
                                           surface can also be called the __(10)__ surface; however, in
     _____ **5.**   four-legged animals, the dorsal surface is the __(11)__
                                           surface.
     _____ **6.**

     _____ **7.**

     _____ **8.**

     _____ **9.**

     _____ **10.**

     _____ **11.**

_____ **12.**

_____ **13.**

_____ **14.**

_____ **15.**

If an incision cuts the heart into right and left parts, the section is a __(12)__ section, but if the heart is cut so that superior and inferior parts result, the section is a __(13)__ section. You are told to cut an animal along two planes so that the paired kidneys are observable in both sections. The two sections that meet this requirement are the __(14)__ and __(15)__ sections.

18. Using key choices, identify the body cavities where the following body organs are located. Enter the appropriate letter or term in the answer blanks.

KEY CHOICES:

A. Abdominopelvic   **B.** Cranial   **C.** Spinal   **D.** Thoracic

_____ **1.** Stomach

_____ **2.** Small intestine

_____ **3.** Large intestine

_____ **4.** Spleen

_____ **5.** Liver

_____ **6.** Spinal cord

_____ **7.** Bladder

_____ **8.** Heart

_____ **9.** Lungs

_____ **10.** Brain

_____ **11.** Rectum

_____ **12.** Ovaries

19. Refer to the organs listed in Exercise 18. In the spaces provided, record the organs that would be found in each of the abdominal regions named here. Some organs may be found in more than one abdominal region.

_____ **1.** Hypogastric region

_____ **2.** Right lumbar region

_____ **3.** Umbilical region

_____ **4.** Epigastric region

_____ **5.** Left iliac region

Basic Chemistry

Everything in the universe is composed of one or more elements, the unique building blocks of all matter. Although over 100 elemental substances exist, only four of these (carbon, hydrogen, oxygen, and nitrogen) make up over 96% of all living material.

The student activities in this chapter test mastery of the basic concepts of both inorganic and organic chemistry. Chemistry is the science that studies the composition of matter. Inorganic chemistry studies the chemical composition of nonliving substances that (generally) do not contain carbon. Organic chemistry studies the carbon-based chemistry (or biochemistry) of living organisms, whether they are maple trees, fish, or humans. Understanding of atomic structure, bonding behavior of elements, and the structure of the most abundant biologic molecules (protein, fats, carbohydrates, and nucleic acids) is tested in various ways to increase the students' understanding of these concepts, which are basic to comprehending the functioning of the body as a whole.

Inorganic Chemistry

1   Define pH in the space provided. _____

_____

2.   Circle the term that does not belong in each of the following groupings.

   1. HCl H_2SO_4 Vinegar   Milk of magnesia

   2. NaOH   Blood (pH 7.4) $AlOH_2$ Urine (pH 6.2)

   3. KCl   NaCl $NaHCO_3$ $MgSO_4$

   4. Ionic   Organic   Salt   Inorganic

   5. pH 7   pH 4   Neutrality $OH^- = H^+$

3. Using key choices, select the appropriate responses to the following descriptive statements. Insert the appropriate letter or term in the answer blanks.

KEY CHOICES:

A. Atom   **D.** Energy   **G.** Molecule   **I.** Protons

B. Electrons   **E.** Ion   **H.** Neutrons   **J.** Valence

C. Element   **F.** Matter

_____   **1.** An electrically charged atom

_____   **2.** Anything that takes up space and has mass (weight)

_____   **3.** A unique, or basic, substance that cannot be decomposed or broken down by ordinary chemical means

_____   **4.** Negatively charged particles, forming part of an atom

_____   **5.** Subatomic particles that determine an atom's chemical behavior, or bonding ability

_____   **6.** The ability to do work

_____   **7.** Building block, or smallest particle, of an elemental substance

_____   **8.** Smallest particle of a substance formed when atoms combine chemically

_____   **9.** Positively charged particles, forming part of an atom

_____   **10.** Name given to the electron shell of an atom that contains the most reactive electrons

_____   **11.** _____   **12.** Subatomic particles responsible for most of an atom's mass.

4. Insert the *chemical symbol* (the chemist's shorthand) in the answer blanks for each of the following elements.

_____   **1.** Oxygen   _____   **7.** Calcium

_____   **2.** Carbon   _____   **8.** Sodium

_____   **3.** Potassium   _____   **9.** Phosphorus

_____   **4.** Iodine   _____   **10.** Magnesium

_____   **5.** Hydrogen   _____   **11.** Chlorine

_____   **6.** Nitrogen   _____   **12.** Iron

5. Match the terms in Column B to the chemical equations listed in Column A.
   Enter the correct letter or term in the answer blanks.

	Column A	Column B
_____	**1.** A + B → AB	**A.** Decomposition
_____	**2.** AB + CD → AD + CB	**B.** Exchange
_____	**3.** XY → X + Y	**C.** Synthesis

6. Figure 2-1 is a diagram of an atom. First select different colors for each of the
   structures listed below. Color in the coding circles and corresponding structures
   on the figure and complete this exercise by responding to the following
   questions, referring to the atom in this figure. Insert your answers in the spaces
   provided.

   ◯ Nucleus

   ◯ Electrons

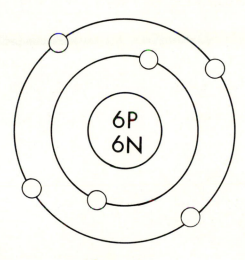

Figure 2-1

   1. What is the atomic number of this atom? _____

   2. What is its atomic mass? _____

   3. What atom is this? _____

   4. If this atom had one additional neutron but the other subatomic particles
      remained the same as shown, this slightly different atom (of the same element)
      would be called a(n) _____

   5. Is this atom chemically active or inert? _____

   6. How many electrons would be needed to fill its outer (valence) shell? _____

7. Would this atom most likely take part in forming ionic or

covalent bonds? _____ Why? _____

_____

7. Both H_2O_2 and $2OH^-$ indicate chemical species with two hydrogen atoms and two oxygen atoms. Briefly explain how these species are different:

_____

_____

8. Two types of chemical bonding are shown in Figure 2-2. In the figure, identify each type as a(n) *ionic* or *covalent* bond. In the case of the ionic bond, indicate which atom has lost an electron by adding a colored arrow to show the direction of electron transfer.

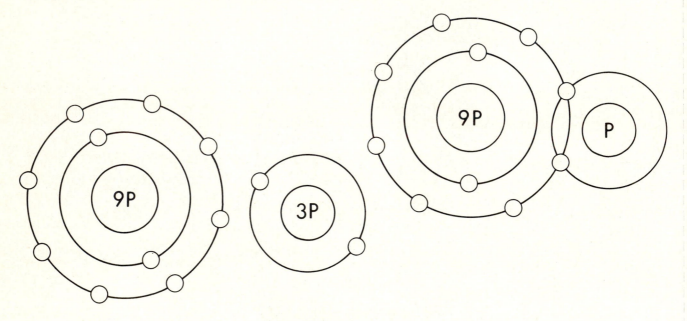

Figure 2-2

9. Use key choices to identify the substances described in the following statements. Insert the appropriate letter or corresponding term in the answer blanks.

KEY CHOICES:

A. Acid(s)   **B.** Base(s)   **C.** Buffer   **D.** Salt

_____ **1.** _____ **2.** _____ **3.** Substances that ionize in water; good electrolytes

_____ **4.** Proton (H^+) acceptor

_____ 5. Ionize in water to release hydrogen ions and a negative ion other than hydroxide (OH⁻)

_____ 6. Ionize in water to release ions other than H^+ and OH^-

_____ 7. Formed when an acid and a base are combined

_____ 8. Substances such as lemon juice and vinegar

_____ 9. Prevents rapid/large swings in pH

10. If the following statements are true, insert *T* in the answer blanks. If any of the statements are false, correct the underlined word(s) and insert your correction in the answer blank.

_____ 1. Na^+ and K^+ are needed for body cells, such as nerve cells, to conduct electrical impulses.

_____ 2. A true solution exhibits the sol–gel phenomenon.

_____ 3. A pH of 1.0 indicates a strongly basic solution.

_____ 4. Water is a polar covalent compound.

_____ 5. A bond formed between two like atoms is always an ionic bond.

_____ 6. Sand and water constitute a solution.

_____ 7. When an atom completely loses an electron to another atom, it becomes an anion.

_____ 8. The atomic number of oxygen is 8. Therefore, oxygen atoms always contain 8 neutrons.

_____ 9. A compound must contain hydrogen to be considered an organic compound.

_____ 10. The universal solute is water.

_____ 11. The greater the distance of an electron from the nucleus, the less energy it has.

_____ 12. Electrons are located in more or less designated areas of space around the nucleus called orbitals.

_____ 13. An unstable atom that decomposes and emits energy is called retroactive.

11. Respond to the following instructions relating to the equation:

$$H_2CO_3 \rightarrow H^+ + HCO_3^-$$

   1. In the space provided, list the chemical formula(s) of compounds. _____

   2. In the space provided, list the chemical formula(s) of ions. _____

   3. Circle the product(s) of the reaction.

   4. Modify the equation by adding a colored arrow in the proper place to indicate that the reaction is reversible.

Organic Chemistry

12. Use an *X* to designate which of the following are organic compounds.

   _____ Nucleic acids   _____ Fats   _____ Proteins

   _____ Oxygen   _____ KCl   _____ Glucose

13. Match the terms in Column B to the descriptions provided in Column A. Enter the correct letter(s) or term(s) in the answer blanks.

Column A	Column B
_____ **1.** Building blocks of carbohydrates	**A.** Amino Acids
_____ **2.** and **3.** Building blocks of fats	**B.** Carbohydrates
_____ **4.** Building blocks of proteins	**C.** Fats (lipids)
_____ **5.** Building blocks of nucleic acids	**D.** Fatty acids
_____ **6.** Cellular cytoplasm is primarily composed of this substance	**E.** Glycerol
	F. Glycogen
_____ **7.** The most important fuel source for body cells	**G.** Nucleotides
_____ **8.** Not soluble in water	**H.** Monosaccharides
_____ **9.** Contains C, H, and O in the ratio CH_2O	**I.** Proteins
_____ **10.** Contain C, H, and O, but have relatively small amounts of oxygen	
_____ **11.** _____ **12.** Contain N in addition to C, H, and O (two answers)	

14. Using key choices, correctly select *all* terms that correspond to the following descriptions. Insert the correct letter(s) or their corresponding term(s) in the answer blanks.

KEY CHOICES:

A. Cholesterol   D. Enzyme   G. Hormones   J. Maltose

B. Collagen   E. Glycogen   H. Keratin   K. RNA

C. DNA   F. Hemoglobin   I. Lactose   L. Starch

_____ 1. Example(s) of structural proteins

_____ 2. Example(s) of functional proteins

_____ 3. Biologic catalyst

_____ 4. Plant storage carbohydrate

_____ 5. Animal storage carbohydrate

_____ 6. The "stuff" of the genes

_____ 7. A steroid

_____ 8. Double sugars, or disaccharides

15. Five simplified diagrams of biologic molecules are depicted in Figure 2-3. First, identify the molecules and insert the correct name in the answer blanks on the figure. Then select a different color for each molecule listed below and use it to color the coding circles and the corresponding molecule on the illustration.

◯ Fat   ◯ Nucleotide   ◯ Monosaccharide

◯ Functional protein   ◯ Polysaccharide

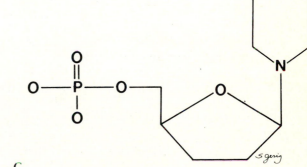

A _____   B _____   C _____

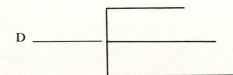

D _____   E _____

Figure 2-3

16. Circle the term that does not belong in each of the following groupings.

    1. Adenine   Guanine   Glucose   Thymine

    2. DNA   Ribose   Phosphate   Deoxyribose

    3. Galactose   Glycogen   Fructose   Glucose

    4. Amino acid   Polypeptide   Glycerol   Protein

    5. Glucose   Sucrose   Lactose   Maltose

    6. Synthesis   Dehydration   Hydrolysis   Water removal

17. Figure 2-4 shows the molecular structure of DNA, a nucleic acid.

A. First, identify the two unnamed nitrogen bases and insert their correct names and symbols in the two blanks beside the color coding circles.
B. Complete the identification of the bases on the diagram by inserting the correct symbols in the appropriate spaces on the right side of the diagram.
C. Select different colors and color the coding circles and the corresponding parts of the diagram.
D. Label one d-R sugar unit and one P unit of the "backbones" of the DNA structure by inserting leader lines and labels on the diagram.
E. Circle the associated nucleotide.

◯ Deoxyribose sugar (d-R)   ◯ Adenine (A)

◯ Phosphate unit (P)   ◯ Cytosine (C)

◯ _____ ( )

◯ _____ ( )

Then answer the following questions by writing your answers in the answer blanks.

1. Name the bonds that help to hold the two DNA strands together. _____

2. Name the three-dimensional shape of the DNA molecule. _____

3. How many base-pairs are present in this segment of a DNA model? _____

4. What is the term that means "base-pairing"?

_____

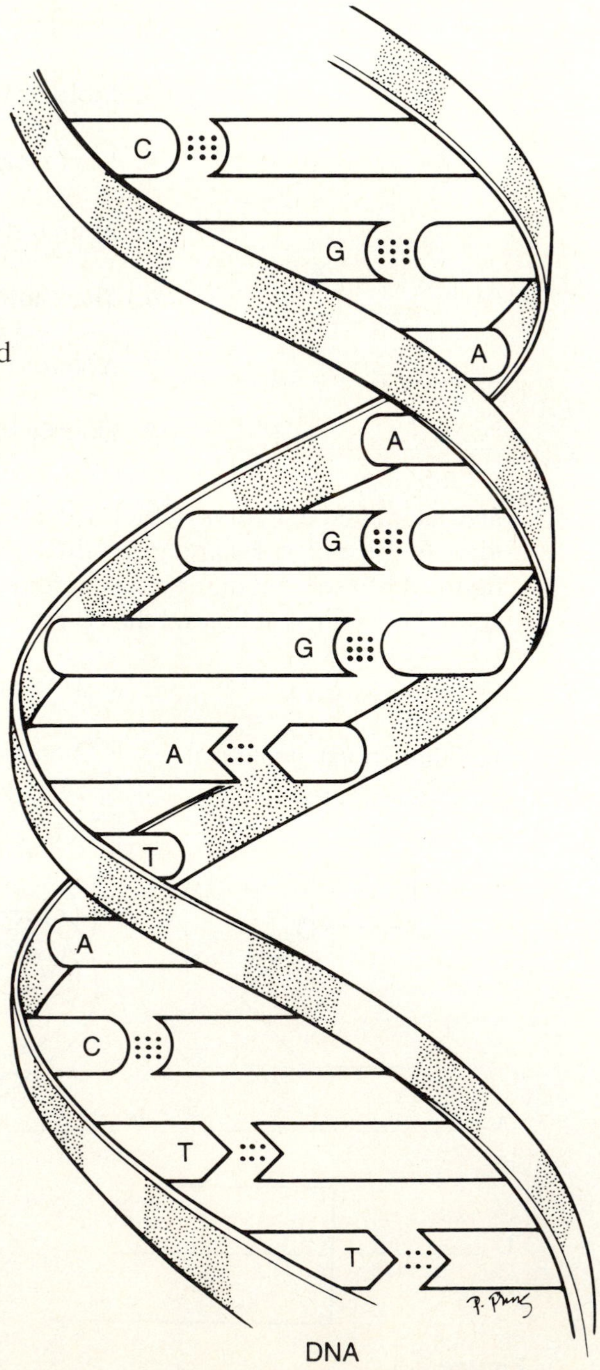

DNA

Figure 2-4: Two DNA strands (coiled)

18. If the following statements are true, insert T in the answer blanks. If any are false, correct the <u>underlined</u> word(s) and insert your correction in the answer blank.

_____ **1.** Phospholipids are <u>polarized</u> molecules.

_____ **2.** <u>Steroids</u> are the major form in which body fat is stored.

_____ **3.** <u>Water</u> is the most abundant compound in the body.

_____ **4.** <u>Nonpolar</u> molecules are generally soluble in water.

_____ **5.** The bases of RNA are A, G, C, and <u>U</u>.

_____ **6.** The universal energy currency of living cells is <u>RNA</u>.

_____ **7.** RNA is <u>single</u> stranded.

_____ **8.** The four elements comprising over 90% of living matter are C, H, N, and <u>Na</u>.

The Incredible Journey:
A Visualization Exercise for
Biochemistry

. . . you are suddenly up-ended and are carried along in a sea of water molecules at almost unbelievable speed.

19. Complete the narrative by inserting the missing words in the answer blanks.

_____ **1.**

_____ **2.**

_____ **3.**

For this journey, you are miniaturized to the size of a very small molecule by colleagues who will remain in contact with you by radio. Your instructions are to play the role of a water molecule and to record any reactions that involve water molecules. Since water molecules are polar molecules, you are outfitted with an insulated rubber wet suit with one __(1)__ charge at your helmet and two __(2)__ charges, one at the end of each leg.

As soon as you are injected into your host's bloodstream, you feel as though you are being pulled apart. Some large, attractive forces are pulling at your legs from different directions! You look about but can see only water molecules. After a moment's thought, you remember the polar nature of your wet suit. You record that these forces must be the __(3)__ in water that are easily formed and easily broken.

_____ 4.

_____ 5.

_____ 6.

_____ 7.

_____ 8.

_____ 9.

_____ 10.

_____ 11.

_____ 12.

_____ 13.

_____ 14.

_____ 15.

After this initial surprise, you are suddenly up-ended and carried along in a sea of water molecules at almost unbelievable speed. You have just begun to observe some huge, red, disk-shaped structures (probably __(4)__) taking up O_2 molecules, when you are swept into a very turbulent environment. Your colleagues radio that you are in the small intestine. With difficulty, because of numerous collisions with other molecules, you begin to record the various types of molecules you see.

In particular, you notice a very long helical molecule made of units with distinctive R-groups. You identify and record this type of molecule as a __(5)__ , made of units called __(6)__ that are joined together by __(7)__ bonds. As you move too close to the helix during your observations, you are nearly pulled apart to form two ions, __(8)__ , but you breathe a sigh of relief as two ions of another water molecule take your place. You watch as these two ions move between two units of the long helical molecule. Then, in a fraction of a second, the bond between the two units is broken. As you record the occurrence of this chemical reaction, called __(9)__ , you are jolted into another direction by an enormous globular protein, the very same __(10)__ that controls and speeds up this chemical reaction.

Once again you find yourself in the bloodstream, heading into an organ identified by your colleagues as the liver. Inside a liver cell, you observe many small monomers, made up only of C, H, and O atoms. You identify these units as __(11)__ molecules because the liver cells are bonding them together to form very long, branched polymers called __(12)__ . You record that this type of chemical reaction is called __(13)__ , and you happily note that this reaction also produces __(14)__ molecules like you!

Via another speedy journey through the bloodstream, you next reach the skin. You move deep into the skin and finally gain access to a sweat gland. In the sweat gland, you collide with millions of water molecules and some ionized salt molecules that are continually attracted to your positive and negative charges. Suddenly, the internal temperature rises, and molecular collisions __(15)__ at an alarming rate, propelling you through the pore of the sweat gland onto the surface of the skin. So that you will be saved from the fate of evaporating into thin air, you contact your colleagues and are speedily rescued.

Cells and Tissues

The basic unit of structure and function in the human body is the cell. Each of a cell's parts, or organelles, as well as the entire cell, is organized to perform a specific function. Cells have the ability to metabolize, grow and reproduce, move, and respond to stimuli. The cells of the body differ in shape, size, and in specific roles in the body. Cells that are similar in structure and function form tissues, which, in turn, form the various body organs.

Student activities in this chapter include questions relating to the structure and function of the generalized animal cell and to the general arrangement of tissues and their contribution to the activities of the various body organs.

Cells

General Understandings

1. Answer the following questions by inserting your responses in the answer blanks.

_____ 1.

_____ 2.

_____ 3.

_____ 4.

_____ 5.

_____ 6.

_____ 7.

_____ 8.

_____ 9.

1–4. Name the four elements that make up the bulk of living matter.

5. Name the single most abundant material or substance in living matter.

6. Name the trace element most important for making bones hard.

7. Name the element, found in small amounts in the body, that is needed to make hemoglobin for oxygen transport.

8–12. Although there are many specific "jobs" that certain cells are able to do, name five functions common to all cells.

_____ 10.

_____ 11.

_____ 12.

_____ 13.

_____ 14.

_____ 15.

_____ 16.

_____ 17.

13–15. List three different shapes of cells.

16. Name the fluid, similar to sea water, that surrounds and bathes all body cells.

17. Name the flattened cells, important in protection, that fit together like tiles. (This is just one example of the generalization that a cell's structure is very closely related to its function in the body.)

Anatomy of a Generalized Animal Cell

2. Complete the following table to fully describe the various cell parts. Insert your responses in the spaces provided under each heading.

Cell structure	Location	Function
	External boundary of the cell	Confines cell contents; regulates entry and exit of materials
Lysosome		
	Scattered throughout the cell	Controls release of energy from foods; forms ATP
	Projections of the cell membrane	Increase the membrane surface area
Golgi apparatus		
Nucleus		
	Two rod-shaped bodies near the nucleus	"Spin" the mitotic spindle
Nucleolus		
Smooth ER		
Rough ER		
	Attached to membrane systems or scattered in the cytoplasm	Synthesize proteins
Chromatin		
Inclusions		

3. Using the following list of terms, correctly label all cell parts indicated by leader lines in Figure 3-1. Then select different colors for each structure and use them to color the coding circles and the corresponding structures in the illustration.

   ○ Plasma membrane   ○ Mitochondrion

   ○ Centriole(s)   ○ Nuclear membrane

   ○ Chromatin thread(s)   ○ Nucleolus

   ○ Golgi apparatus   ○ Rough endoplasmic reticulum (ER)

   ○ Microvilli   ○ Smooth endoplasmic reticulum (ER)

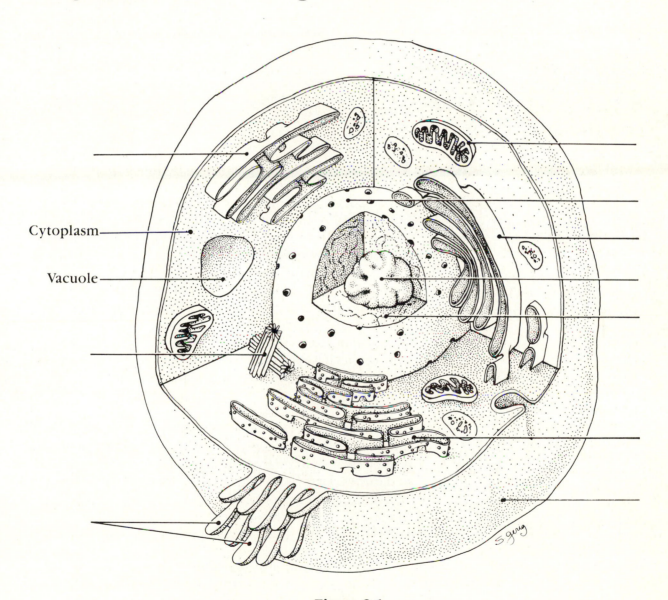

Cytoplasm

Vacuole

Figure 3-1

Cell Physiology

4. The following statements refer specifically to DNA (genetic material) and its role in the body. Choose responses from the key choices that complete these statements. Insert the appropriate letter or term in the answer blanks.

KEY CHOICES:

A. Adenine	**H.** Enzyme(s)	**O.** Old	**U.** Sugar
B. Amino acid(s)	**I.** Functional proteins	**P.** Phosphate	**V.** Structural protein(s)
C. Bases	**J.** Genes	**Q.** Protein	**W.** Template, or model
D. Codons	**K.** Growth	**R.** Replication	**X.** Thymine
E. Complementary	**L.** Guanine	**S.** Repair	**Y.** Transcription
F. Cytosine	**M.** New	**T.** Ribosome	**Z.** Uracil
G. Double helix	**N.** Nucleotides		

_____ 1.

_____ 2.

_____ 3.

_____ 4.

_____ 5.

_____ 6.

_____ 7.

_____ 8.

_____ 9.

_____ 10.

_____ 11.

_____ 12.

_____ 13.

_____ 14.

_____ 15.

DNA molecules contain information for building specific __(1)__ . In a three-dimensional view, a DNA molecule looks like a spiral staircase; this is correctly called a __(2)__ . The constant parts of DNA molecules are the __(3)__ and __(4)__ molecules, forming the DNA-ladder uprights or backbones. The information of DNA is actually coded in the sequence of nitrogen-containing __(5)__ , which is (are) bound together to form the "rungs" of the DNA ladder. When the four DNA bases are combined in different three-base sequences, different __(6)__ of the protein are called for. It is said that the N-containing bases of DNA are __(7)__ , which means that only certain bases can fit or interact together. Specifically this means that guanine can bind with __(8)__ , and adenine binds with __(9)__ .

The production of proteins involves the cooperation of DNA and RNA. RNA is another type of nucleic acid that serves as a "molecular slave" to DNA. That is, it leaves the nucleus and carries out the instructions of the DNA for the building of a protein on a cytoplasmic structure called a __(10)__ . There are two major types of proteins: __(11)__ , which build parts of cells, and __(12)__ , which perform other functions for cells such as carrying oxygen, regulating growth, or helping to regulate metabolism. A very important group of functional proteins is the __(13)__ , which act as biochemical catalysts. These catalysts increase the rate of chemical reactions without becoming part of the products.

_____ 16.

_____ 17.

_____ 18.

_____ 19.

_____ 20.

_____ 21.

_____ 22.

When a cell is preparing to divide, in order for its daughter cells to have all its information, it must oversee the __(14)__ of its DNA so that a "double dose" of genes is present for a brief period. For DNA synthesis to occur, the DNA must uncoil and the bonds between the N-bases must be broken. Then the two single strands of __(15)__ each act as a __(16)__ for the building of a whole DNA molecule. When completed, each DNA molecule formed is half __(17)__ and half __(18)__. The fact that DNA replicates before a cell divides ensures that each daughter cell has a complete set of __(19)__. Cell division, which then follows, provides new cells so that __(20)__ and __(21)__ can occur.

5. Complete the following statements that relate to the transport of substances through cellular membranes. For each statement, choose all answers that apply and place their letters in the answer blanks.

1. Kinetic energy _____.

   A. is higher in larger molecules.

   B. is lower in larger molecules.

   C. increases with increasing temperature.

   D. decreases with increasing temperature.

   E. is shown in the speed of molecular movement.

2. Selective permeability _____.

   A. means that some substances can pass passively, but others must be actively transported.

   B. means that the membrane is permeable to everything.

   C. means that the membrane is impermeable to everything.

   D. can occur only in _living_ cells.

   E. is important in obtaining nutrients and getting rid of wastes.

6. A semipermeable sac, containing 4% NaCl, 9% glucose, and 10% albumin, is suspended in a solution with the following composition: 10% NaCl, 10% glucose, and 40% albumin. Assume the sac is permeable to all substances *except* albumin. Using the key choices, insert the letter indicating the correct event in the answer blanks.

KEY CHOICES:

A. Moves into the sac   **B.** Moves out of the sac   **C.** Does not move

_____   **1.** Glucose

_____   **2.** Water

_____   **3.** Albumin

_____   **4.** NaCl

7. Figure 3-2 shows three microscopic fields (A–C) containing red blood cells. Arrows indicate the direction of net osmosis. Respond to the following questions, referring to Figure 3-2, by inserting your responses in the spaces provided.

   1. Which microscopic field contains a *hypertonic* solution? _____

   The cells in this field are said to be_____

   2. Which microscopic field contains an isotonic bathing solution? _____

   What does *isotonic* mean? _____

   3. Which microscopic field contains a *hypotonic* solution? _____

   What is happening to the cells in this field? _____

   Why? _____

   _____

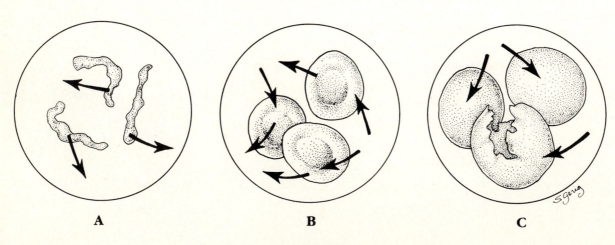

A   B   C

Figure 3-2

8. Select the key choices that characterize each of the following statements. Insert the appropriate letter(s) or corresponding terms in the answer blanks.

KEY CHOICES:

A. Diffusion, dialysis	**C.** Filtration	**E.** Pinocytosis
B. Diffusion, osmosis	**D.** Phagocytosis	**F.** Solute pumping

_____ **1.** Require ATP (cellular energy)

_____ **2.** Driven by kinetic energy of the molecules

_____ **3.** Driven by hydrostatic (fluid) pressure

_____ **4.** Follow a concentration gradient

_____ **5.** Proceeds against a concentration gradient; require(s) a carrier

_____ **6.** Engulf foreign substances

_____ **7.** Moves water through a semipermeable membrane

_____ **8.** Transports amino acids, some sugars, and Na^+ through the plasma membrane

_____ **9.** Provides for cellular uptake of solid or large particles from the cell exterior

_____ **10.** Moves small or lipid-soluble solutes through the membrane

9. Figure 3-3 is a diagrammatic representation of a portion of a plasma membrane. Select two different colors for lipid and protein molecules. Color the coding circles and the corresponding molecules in the illustration. Then add a colored arrow for each substance shown inside and outside the cell indicating (a) its *direction* of transport through the membrane; and (b) its *means of transport* (that is, either directly through the lipid portion or by attachment to a protein carrier).

◯ Lipid molecules

◯ Protein molecules

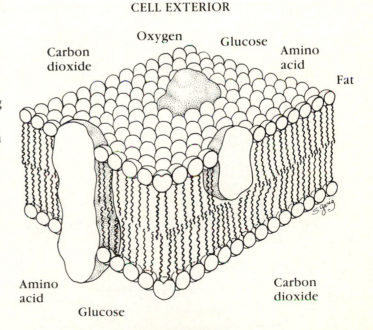

CELL EXTERIOR

CELL INTERIOR

Figure 3-3

Cell Division

10. Identify the phases of mitosis depicted in Figure 3-4 by inserting the correct name in the blank under the appropriate diagram. Then select different colors to represent the structures listed below and use them to color in the coding circles and the corresponding structures in the illustration.

◯ Nuclear membrane(s), if present   ◯ Centrioles

◯ Nucleoli, if present   ◯ Spindle fibers

◯ Chromosomes

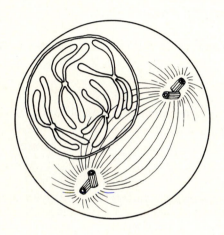

A _____

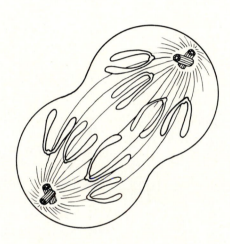

B _____

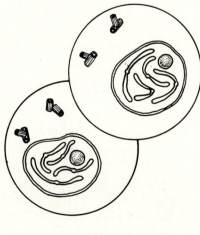

C _____

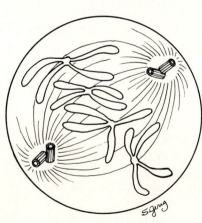

D _____

Figure 3-4

11. The following statements describe events that occur during the different phases of mitosis. Identify the phase by choosing the correct response(s) from key choices and inserting the letter(s) or term(s) in the answer blanks.

KEY CHOICES:

A. Anaphase   **C.** Prophase   **E.** None of these

B. Metaphase   **D.** Telophase

_____ **1.** Chromatin coils and condenses to form deeply staining bodies.

_____ **2.** Centromeres break, and chromosomes begin migration toward opposite poles of the cell.

_____ **3.** The nuclear membrane and nucleoli reappear.

_____ **4.** When chromosomes cease their poleward movement, this phase begins.

_____ **5.** Chromosomes align on the equator of the spindle.

_____ **6.** The nucleoli and nuclear membrane disappear.

_____ **7.** The spindle forms through the migration of the centrioles.

_____ **8.** Chromosomal material replicates.

_____ **9.** Chromosomes first appear to be duplex structures.

_____ **10.** Chromosomes attach to the spindle fibers.

_____ **11.** A cleavage furrow forms.

_____ **12.** The nuclear membrane is absent during the entire phase.

_____ **13.** Period during which a cell carries out its *usual* metabolic activities.

12. Complete the following statements. Insert your answers in the answer blanks.

_____ **1.**

_____ **2.**

_____ **3.**

_____ **4.**

_____ **5.**

_____ **6.**

_____ **7.**

Division of the __(1)__ is referred to as mitosis. Cytokinesis is division of the __(2)__. The major structural difference between chromatin and chromosomes is that the latter is __(3)__. Chromosomes attach to the spindle fibers by undivided structures called __(4)__. If a cell undergoes nuclear division but not cytoplasmic division, the product is a __(5)__. The structure that acts as a scaffolding for chromosomal attachment and movement is called the __(6)__. __(7)__ is the period of cell life when the cell is not involved in division.

Tissues of the Body

13. Twelve tissue types are diagrammed in Figure 3-5. Identify each tissue type by inserting the correct name in the blank below it on the diagram. Select different colors for the following structures and use them to color the coding circle and corresponding structures in the diagrams.

- ◯ Epithelial cells
- ◯ Muscle cells
- ◯ Nerve cells
- ◯ Matrix (Where found, matrix should be colored differently from the living cells of that tissue type. Be careful, this may not be as easy as it seems!)

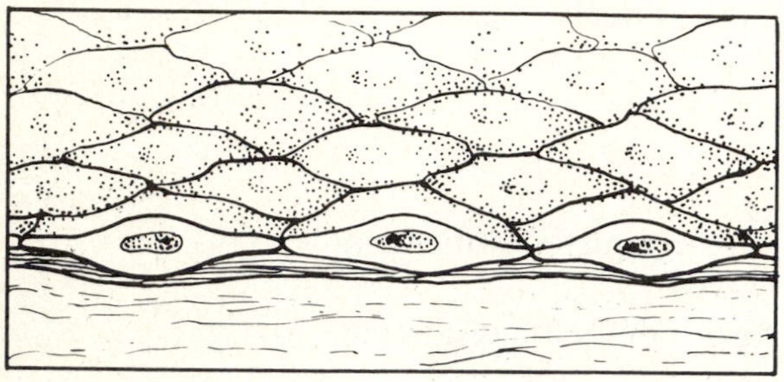

A _____

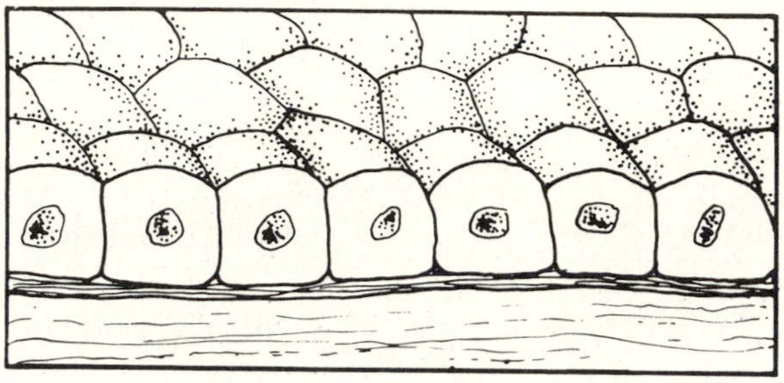

B _____

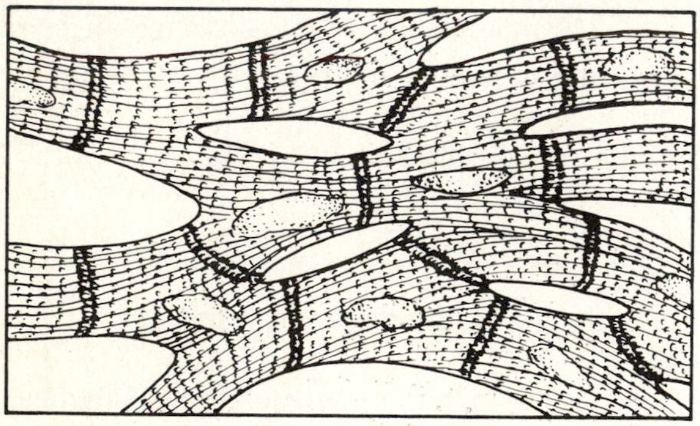

C _____

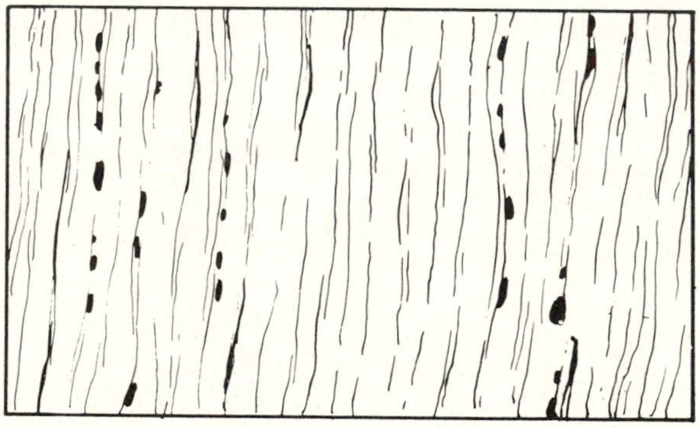

D _____

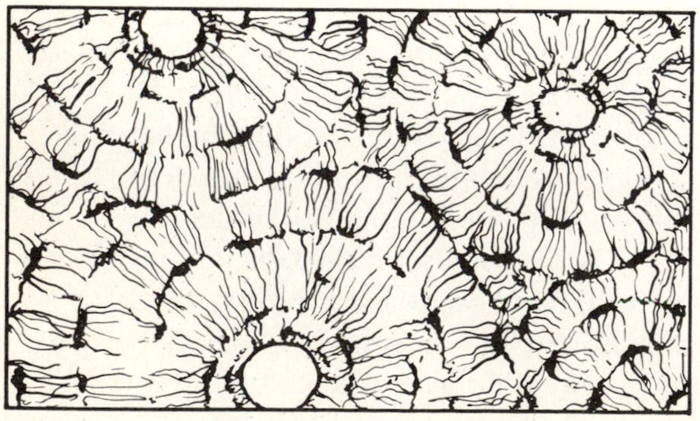

E _____

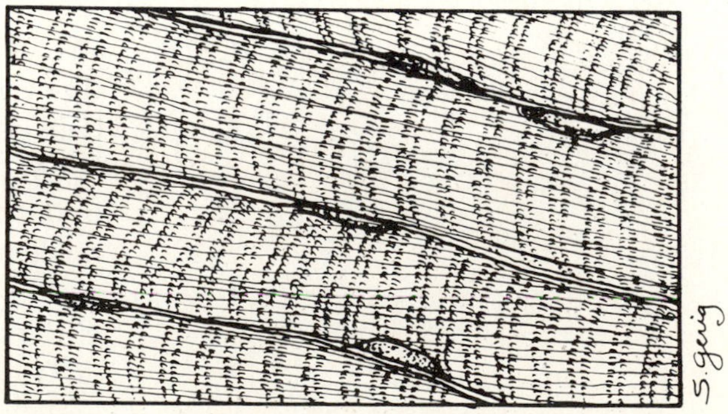

F _____

Figure 3-5, A-F

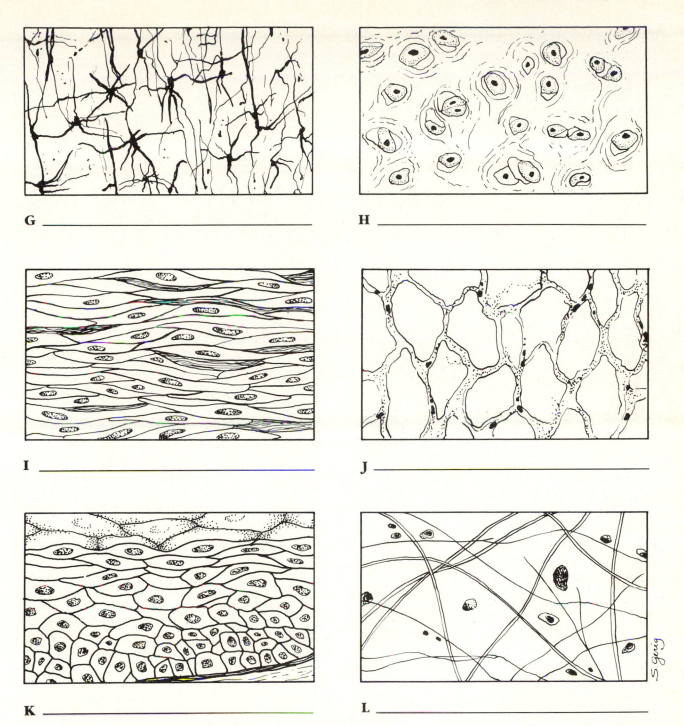

G _____

H _____

I _____

J _____

K _____

L _____

Figure 3-5, G-L

14. Describe briefly how the particular structure of a neuron relates to its function

    in the body. _____

    _____

    _____

15. Using key choices, correctly identify the *major* tissue types described. Enter the appropriate letter or tissue type term in the answer blanks.

KEY CHOICES:

A. Connective   B. Epithelium   C. Muscle   D. Nervous

_____ 1. Forms mucous, serous, and epidermal membranes

_____ 2. Allows for movement of limbs and for organ movements within the body

_____ 3. Transmits electrochemical impulses

_____ 4. Supports body organs

_____ 5. Cells of this tissue may absorb and/or secrete substances

_____ 6. Basis of the major controlling system of the body

_____ 7. The major function of the cells of this tissue type is to shorten

_____ 8. Forms hormones

_____ 9. Packages and protects body organs

_____ 10. Characterized by having large amounts of nonliving matrix

_____ 11. Allows you to smile, grasp, swim, ski, and shoot an arrow

_____ 12. Most widely distributed tissue type in the body

_____ 13. Forms the brain and spinal cord

16. Using key choices, identify the following specific type(s) of epithelial tissue. Enter the appropriate letter or classification term in the answer blanks.

KEY CHOICES:

A. Pseudostratified ciliated   C. Simple cuboidal   E. Stratified squamous

B. Simple columnar   D. Simple squamous   F. Transitional

_____ 1. Forms the lining of the esophagus

_____ 2. Forms the lining of the stomach and small intestine

_____ 3. Found in lung tissue (alveolar sacs)

_____ 4. Forms the collecting tubules of the kidney

_____ 5. Forms the epidermis of the skin

_____ 6. Found in the bladder lining; peculiar cells that slide over one another

_____ 7. Forms thin serous membranes; a single layer of flattened cells

17. The three types of muscle tissue exhibit certain similarities and differences. Check (√) the appropriate spaces in the following table to indicate which muscle types exhibit each characteristic

Characteristic	Skeletal	Cardiac	Smooth
1. Voluntarily controlled			
2. Involuntarily controlled			
3. Banded appearance			
4. Single nucleus in each cell			
5. Multinucleate			
6. Found attached to bones			
7. Allows you to direct your eyeballs			
8. Found in the walls of stomach, uterus, and arteries			
9. Contains spindle-shaped cells			
10. Contains cylindrical cells with branching ends			
11. Contains long, nonbranching cylindrical cells			
12. Displays intercalated disks			
13. Concerned with locomotion of the body as a whole			
14. Changes the internal volume of an organ as it contracts			
15. Tissue of the circulatory pump			

18. Circle the term that does not belong in each of the following groupings.

1. Columnar   Areolar   Cuboidal   Squamous

2. Collagen   Cell   Matrix   Cell product

3. Cilia   Flagellum   Microvilli   Elastic fibers

4. Glands   Bones   Epidermis   Mucosae

5. Adipose   Hyaline   Osseous   Nervous

6. Blood   Smooth   Cardiac   Skeletal

19. Using key choices, identify the following connective tissue types. Insert the appropriate letter or corresponding term in the answer blanks.

KEY CHOICES:

A. Adipose connective tissue   **D.** Osseous tissue

B. Areolar connective tissue   **E.** Hemopoietic tissue

C. Dense fibrous connective tissue   **F.** Hyaline cartilage

_____ **1.** Provides great strength through parallel bundles of collagenic fibers; found in tendons

_____ **2.** Acts as a storage depot for fat

_____ **3.** Composes the dermis of the skin

_____ **4.** Forms the bony skeleton

_____ **5.** Composes the basement membrane and packages organs; includes a gel-like matrix with all categories of fibers and many cell types

_____ **6.** Forms the embryonic skeleton and the surfaces of bones at the joints; reinforces the trachea

_____ **7.** Provides insulation for the body

_____ **8.** Structurally amorphous matrix, heavily invaded with fibers; appears glassy and smooth

_____ **9.** Contains cells arranged concentrically around a nutrient canal; matrix is hard due to calcium salts

Tissue Repair

20. For each of the following statements about tissue repair that is true, enter *T* in the answer blank. For each false statement, correct the underlined words by writing the correct words in the answer blank.

_____ **1.** The nonspecific response of the body to injury is called underline{regeneration}.

_____ **2.** Intact capillaries near an injury dilate, leaking plasma, blood cells, and underline{antibodies}, which cause the blood to clot. The clot at the surface dries to form a scab.

_____ **3.** During the first phase of tissue repair, capillary buds invade the clot, forming a delicate pink tissue called underline{endodermal} tissue.

_____ **4.** When damage is not too severe, the surface epithelium migrates beneath the dry scab and across the surface of the granulation tissue. This repair process is called underline{proliferation}.

_____ **5.** If tissue damage is very severe, tissue repair is more likely to occur by <u>fibrosis</u>, or scarring.

_____ **6.** During fibrosis, fibroblasts in the granulation tissue lay down <u>keratin</u> fibers, which form a strong, compact, but inflexible mass.

_____ **7.** The repair of cardiac muscle and nervous tissue occurs only by <u>fibrosis</u>.

Developmental Aspects of Cells and Tissues

21. Correctly complete each statement by inserting your responses in the answer blanks.

_____ **1.**

_____ **2.**

_____ **3.**

_____ **4.**

_____ **5.**

_____ **6.**

_____ **7.**

_____ **8.**

_____ **9.**

_____ **10.**

_____ **11.**

_____ **12.**

_____ **13.**

_____ **14.**

_____ **15.**

_____ **16.**

_____ **17.**

_____ **18.**

_____ **19.**

_____ **20.**

During embryonic development, cells specialize to form __(1)__. Mitotic cell division is very important for overall body __(2)__. All tissues except __(3)__ tissue continue to undergo cell division until the end of adolescence. After this time, __(4)__ tissue also becomes amitotic. When amitotic tissues are damaged, they are replaced by __(5)__ tissue, which does not function in the same way as the original tissue. This is a serious problem when heart cells are damaged.

Aging begins almost as soon as we are born. Three explanations of the aging process have been offered. One states that __(6)__ insults such as the presence of toxic substances in the blood are important; another theory states that external __(7)__ factors such as X rays help to cause aging; a third theory suggests that aging is programmed in our __(8)__. Three examples of aging processes seen in all people are __(9)__, __(10)__, and __(11)__.

Neoplasms occur when cells "go wild" and the normal controls of cell __(12)__ are lost. The two types of neoplasms are __(13)__ and __(14)__. The __(15)__ type tends to stay localized and have a capsule; the __(16)__ type is likely to invade other body tissues and spread to other (distant) parts of the body. To correctly diagnose the type of neoplasm, a microscopic examination of the tissue called a __(17)__ is usually done. Whenever possible, __(18)__ is the treatment of choice for neoplasms.

An overgrowth of tissue that is *not* considered to be a neoplasm is referred to as __(19)__. Conversely, a decrease in the size of an organ or tissue, resulting from loss of normal stimulation, is called __(20)__.

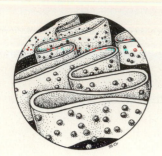

The Incredible Journey:
A Visualization Exercise for the Cell

A long, meandering membrane with dark globules clinging to its outer surface now comes into sight.

22. Where necessary, complete statements by inserting the missing words in the answer blanks.

_____ 1.

_____ 2.

_____ 3.

_____ 4.

_____ 5.

_____ 6.

_____ 7.

_____ 8.

_____ 9.

_____ 10.

_____ 11.

_____ 12.

For your second journey, you will be miniaturized to the size of a small protein molecule and will travel in a microsubmarine, specially designed to enable you to pass easily through living membranes. You are injected into the intercellular space between two epithelial cells, and you are instructed to observe one of these cells firsthand and to identify as many of its structures as possible.

You struggle briefly with the controls and then maneuver your microsub into one of these cells. Once inside the cell, you find yourself in a kind of "sea." This salty fluid that surrounds you is the __(1)__ of the cell.

Far below looms a large, dark oval structure, much larger than anything else you can see. You conclude that it is the __(2)__. As you move downward, you pass a cigar-shaped structure with strange-looking folds on its inner surface. Although you have a pretty good idea that it must be a __(3)__, you decide to investigate more thoroughly. After passing through the external membrane of the structure, you are confronted with yet another membrane. Once past this membrane, you are inside the strange-looking structure. You activate the analyzer switch in your microsub for a readout indicating which molecules are in your immediate vicinity.

As suspected, there is an abundance of energy-rich __(4)__ molecules. Having satisfied your curiosity, you leave this structure to continue the investigation.

A long, meandering membrane with dark globules clinging to its outer surface now comes into sight. You maneuver closer and sit back to watch the activity. As you watch, amino acids are joined together and a long, threadlike protein molecule is built. The globules must be __(5)__, and the membrane, therefore, is the __(6)__. Once again you head toward the large dark structure seen and tentatively identified earlier. On approach, you observe that this huge structure has very large openings in its outer wall; these openings must be the __(7)__. Passing through one of these openings, you discover that from the inside the color of this structure is a result of dark, coiled, intertwined masses of __(8)__, which your analyzer confirms contain genetic material, or __(9)__ molecules. Making your way through this tangled mass, you pass two round, dense structures that appear to be full of the same type of globules you saw outside. These two round structures are __(10)__. All this information confirms your earlier identification of this cellular structure, so now you move to its exterior to continue observations.

Just ahead, you see what appears to be a mountain of flattened sacs with hundreds of small sac-like vesicles at its edges. The vesicles appear to be migrating away from this area and heading toward the outer edges of the cell. The mountain of sacs must be the __(11)__ . Eventually you come upon a rather simple-looking membrane-bound sac. Although it doesn't look too exciting, and has few distinguishing marks, it does not resemble anything else you have seen so far. Deciding to obtain a chemical analysis before entering this sac, you activate the analyzer and on the screen you see "Enzymes — Enzymes — Danger — Danger." There is little doubt that this innocent-appearing structure is actually a __(12)__ .

Completing your journey, you count the number of organelles identified so far. Satisfied that you have observed most of them, you request retrieval from the intercellular space.

Skin and Body Membranes

Body membranes, which cover surfaces, line body cavities, and form protective sheets around organs, fall into two major categories. These are epithelial membranes (skin epidermis, mucosae, and serosae) and the connective tissue synovial membranes.

Topics for review in this chapter include a comparison of structure and function of various membranes, anatomic characteristics of the skin (composed of the connective tissue dermis and the epidermis) and its derivatives, and the manner in which the skin responds to both internal and external stimuli to protect the body.

Integumentary System (Skin)

Basic Structure and Function

1. Complete the following statements in the blanks provided.

   _____ 1. Radiation from the skin surface and evaporation of sweat are two ways in which the skin helps to get rid of body __(1)__ .

   _____ 2. Fat in the __(2)__ tissue layer beneath the dermis helps to insulate the body.

   _____ 3. The waterproofing protein found in the epidermal cells is called __(3)__ .

   _____ 4. A vitamin that is manufactured in the skin is __(4)__ .

   _____ 5. A localized concentration of melanin is __(5)__ .

   _____ 6. Wrinkling of the skin is due to loss of the __(6)__ of the skin.

   _____ 7. A decubitus ulcer results when skin cells are deprived of __(7)__ .

   _____ 8. __(8)__ is a bluish cast of the skin resulting from inadequate oxygenation of the blood.

2. Figure 4-1 is a diagram of the skin. Label the skin structures and areas indicated by leader lines on the figure. Select different colors for the structures listed below and color the coding circles and corresponding structures on the figure.

◯ Arrector pili muscle   ◯ Hair follicle

◯ Adipose tissue   ◯ Sudoriferous gland

◯ Hair bulb   ◯ Sebaceous gland

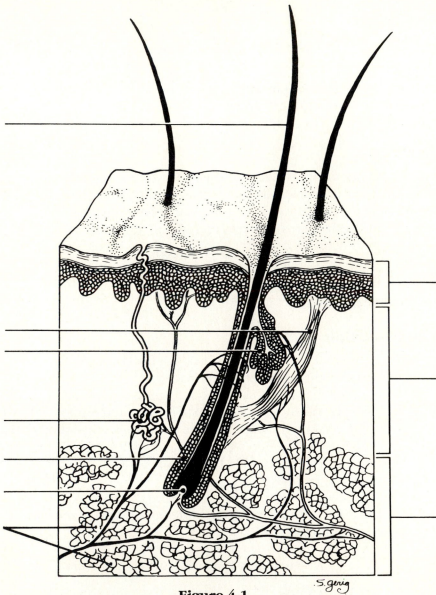

Figure 4-1

3. Using key choices, choose all responses that apply to the following descriptions. Enter the appropriate letter(s) or terms in the answer blanks.

    KEY CHOICES:

    A. Stratum corneum   **D.** Stratum lucidum   **G.** Epidermis as a whole

    B. Stratum germinativum   **E.** Papillary layer   **H.** Dermis as a whole

    C. Stratum granulosum   **F.** Reticular layer

    _____ **1.** Translucent cells, containing keratin

    _____ **2.** Dead cells

    _____ **3.** Dermis layer responsible for fingerprints

    _____ **4.** Vascular region

    _____ **5.** Epidermal region involved in rapid cell division; most inferior epidermal layer

    _____ **6.** Scalelike cells full of keratin that constantly flake off

    _____ **7.** Site of elastic and collagen fibers

    _____ **8.** Site of melanin formation

    _____ **9.** Major skin area from which the derivatives (hair, nails) arise

4. This section reviews burns. Using key choices, select the burn type identified by the following observations. Enter the correct letter or corresponding term in the answer blanks.

    KEY CHOICES:

    A. First-degree burn   **B.** Second-degree burn   **C.** Third-degree burn

    _____ **1.** Full-thickness burn; epidermal and dermal layers destroyed; skin is blanched

    _____ **2.** Blisters form

    _____ **3.** Epidermal damage, redness, and some pain (usually brief)

    _____ **4.** Epidermal and some dermal damage; pain, but regeneration is possible

    _____ **5.** Regeneration impossible; requires grafting

    _____ **6.** Pain absent, because nerve endings in the area are destroyed

5. Name the two major problems resulting from a severe burn:

_____ and _____.

Appendages

6. Using key choices, complete the following statements. Insert the appropriate
letter or term in the answer blanks.

KEY CHOICES:

A. Arrector pili   **D.** Hair follicle(s)   **F.** Sudoriferous gland (apocrine)

B. Cutaneous receptors   **E.** Sebaceous glands   **G.** Sudoriferous gland (eccrine)

C. Hair

_____   **1.** A blackhead is an accumulation of oily material produced by __(1)__ .

_____   **2.** Tiny muscles attached to hair follicles that pull the hair upright during fright or cold are called __(2)__ .

_____   **3.** The most numerous variety of perspiration gland is the __(3)__ .

_____   **4.** A sheath formed of both epithelial and connective tissues is the __(4)__ .

_____   **5.** A less numerous variety of perspiration gland is the __(5)__ . Its secretion (often milky in appearance) contains proteins and other substances that favor bacterial growth.

_____   **6.** __(6)__ is found everywhere on the body except the palms of the hands, soles of the feet, and lips and primarily consists of dead keratinized cells.

_____   **7.** __(7)__ are specialized nerve endings that respond to temperature and touch, for example.

_____   **8.** __(8)__ become more active at puberty.

_____   **9.** Part of the heat-liberating apparatus of the body is the __(9)__ .

7. Circle the term that does not belong in each of the following groupings.

   1. Sebaceous gland   Hair   Arrector pili   Sweat gland

   2. Radiation   Absorption   Conduction   Evaporation

   3. Stratum corneum   Nails   Hair   Stratum germinativum

   4. Freckles   Blackheads   Moles   Melanin

   5. Scent glands   Eccrine glands   Apocrine glands   Axilla

8. If the following statements are true, insert *T* in the answer blank. If any of the
statements are false, correct the <u>underlined</u> word(s) and insert your correction
in the answer blank.

_____ **1.** Greater amounts of the pigment <u>carotene</u> are produced when the
skin is exposed to the sun.

_____ **2.** The most abundant protein in dead epidermal structures such as
hair and nails is <u>melanin</u>.

_____ **3.** <u>Sebum</u> is an oily mixture of lipids, cholesterol, and cell fragments.

_____ **4.** The oldest epidermal cells in the epidermis are found in the
<u>stratum germinativum</u>.

_____ **5.** The externally observable part of a hair is called the <u>root</u>.

_____ **6.** The <u>epidermis</u> provides mechanical strength to the skin.

9. Figure 4-2 is a diagram of a cross-sectional view of a hair in its follicle.
Complete this figure by following the directions in steps 1–3.

1. Identify the two portions of the follicle wall by placing the correct name of the sheath at the
end of the appropriate leader line.

2. Use different colors to color these regions.

3. Label, color code, and color the three following regions of the hair.

◯ Cortex   ◯ Cuticle   ◯ Medulla

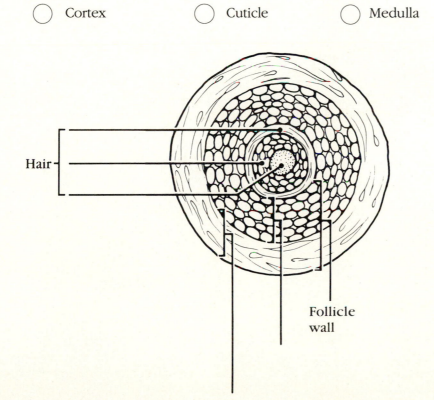

Hair

Follicle
wall

Figure 4-2

Classification of Body Membranes

10. Complete the following table relating to body membranes. Enter your responses in the areas left blank.

Membrane	Tissue type (epithelial/connective)	Common locations	Functions
Mucous			
Serous			
Cutaneous			
Synovial			

11. Five simplified diagrams are shown in Figure 4-3. Select different colors for the membranes listed below, and use them to color the coding circles and the corresponding structures.

○ Mucosae

○ Visceral pleura (serosa)

○ Parietal pleura (serosa)

○ Visceral pericardium (serosa)

○ Parietal peritoneum (serosa)

○ Visceral peritoneum (serosa)

○ Mesentery

○ Endothelium

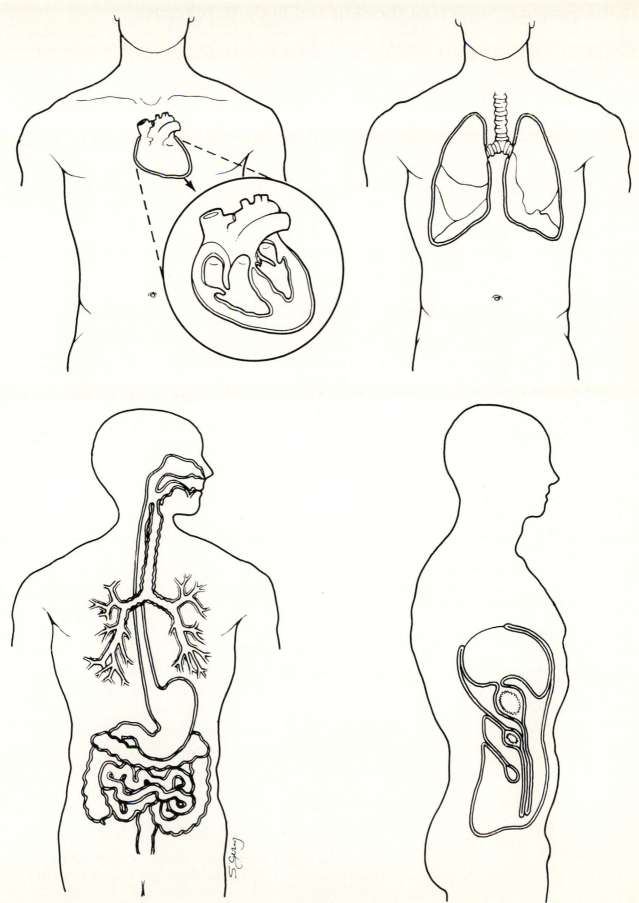

Figure 4-3

Developmental Aspects of the Skin and Body Membranes

12. Match the choices (letters or terms) in Column B with the appropriate descriptions in Column A

Column A

_____ **1.** Skin inflammations that increase in frequency with age

_____ **2.** Cause of graying hair

_____ **3.** Small white bumps on the skin of newborn babies, resulting from accumulations of sebaceous gland material

_____ **4.** Reflects the loss of insulating subcutaneous tissue with age

_____ **5.** A common consequence of accelerated sebaceous gland activity during adolescence

_____ **6.** Oily substance produced by the fetus's sebaceous glands

_____ **7.** The hairy "cloak" of the fetus

Column B

A. Acne

B. Cold intolerance

C. Dermatitis

D. Delayed action gene

E. Lanugo

F. Milia

G. Vernix caseosa

13. Name three factors, other than the factor identified in Exercise 12, that can cause hair loss.

_____ , _____ , _____

The Incredible Journey: A Visualization Exercise for the Skin

Your immediate surroundings resemble huge grotesquely twisted vines. . . . you begin to climb upward.

14. Where necessary, complete statements by inserting the missing words in the answer blanks.

_____ **1.**

_____ **2.**

For this trip, you are miniaturized for injection into your host's skin. Your journey begins when you are injected into a soft gel-like substance. Your immediate surroundings

3. _____

4. _____

5. _____

6. _____

7. _____

8. _____

9. _____

10. _____

resemble huge grotesquely twisted vines. But when you peer carefully at the closest "vine," you realize you are actually seeing connective tissue fibers. Although tangled together, most of the fibers are fairly straight and look like strong cables. You identify these as the __(1)__ fibers. Here and there are fibers that resemble coiled springs. These must be the __(2)__ fibers that help to give skin its springiness. At this point, there is little question that you are in the __(3)__ region of the skin, particularly since you can also see blood vessels and nerve fibers around you.

Carefully, using the fibers as steps, you begin to climb upward. After climbing for some time and finding that you still haven't reached the upper regions of the skin, you stop for a rest. As you sit, a strange-looking cell approaches, moving slowly with parts alternately flowing forward and then receding. Suddenly you realize that this must be a __(4)__ that is about to dispose of an intruder (you) unless you move in a hurry! You scramble to your feet and resume your upward climb. On your right is a large fibrous structure that looks like a tree trunk anchored in place by muscle fibers. By scurrying up this __(5)__ sheath, you are able to escape from the cell. Once safely out of harm's way, you again scan your surroundings. Directly overhead are tall cubelike cells, forming a continuous sheet. In your rush to escape you have reached the __(6)__ layer of the skin. As you watch the activity of the cells in this layer, you notice that many of the cells are pinching in two, and that the daughter cells are being forced upward. Obviously, this is the basal layer that continually replaces cells that rub off the skin surface, and these cells are the __(7)__ cells.

Looking through the transparent cell membrane of one of the basal cells, you see a dark mass hanging over its nucleus. You wonder if this cell could have a tumor; but then, looking through the membranes of the neighboring cells, you find that they also have dark umbrella-like masses hanging over their nuclei. As you consider this matter, a black cell with long tentacles begins to pick its way carefully between the other cells. As you watch with interest, one of the transparent cells engulfs the tip of one of the black cell's tentacles. Within seconds a black substance appears above the transparent cell's nucleus. Suddenly, you remember that one of the skin's functions is to protect the deeper layers from sun damage; the black substance must be the protective pigment, __(8)__ .

Once again you begin your upward climb and notice that the cells are becoming shorter and harder and are full of a waxy substance. This substance has to be __(9)__ , which would account for the increasing hardness of the cells. Climbing still higher, the cells become flattened like huge shingles. The only material apparent in the cells is the waxy substance—there is no nucleus, and there appears to be no activity in these cells. Considering the clues—shinglelike cells, no nuclei, full of the waxy substance, no activity—these cells are obviously __(10)__ and therefore very close to the skin surface.

Suddenly, you feel a strong agitation in your immediate area. The pressure is tremendous. Looking upward through the transparent cell layers, you see your host's fingertips vigorously scratching the area directly overhead. You wonder if you are causing his skin to sting or tickle. Then, within seconds, the cells around you begin to separate and fall apart, and you are catapulted out into the sunlight. Since the scratching fingers might descend once again, you quickly advise your host of your whereabouts.

Skeletal System

The skeleton is constructed of two of the most supportive tissues found in the human body—cartilage and bone. Besides supporting and protecting the body as an internal framework, the skeleton also provides a system of levers that work with skeletal muscles to move the body. In addition, the bones provide a storage depot for such substances as lipids and calcium, and blood cell formation goes on within their red marrow cavities.

The skeleton consists of bones connected at joints, or articulations, and is subdivided into two divisions. The axial skeleton includes those bones that lie around the body's center of gravity, and the appendicular skeleton includes the bones of the limbs.

Topics for student review include structure and function of long bones, location and naming of specific bones in the skeleton, fracture types, and a classification of joint types in the body.

Bones—An Overview

1. Classify each of the following terms as a projection (*P*) or a depression or opening (*D*). Enter the appropriate letter in the answer blanks.

   ____ **1.** Condyle   ____ **4.** Foramen   ____ **7.** Ramus

   ____ **2.** Crest   ____ **5.** Head   ____ **8.** Spine

   ____ **3.** Fissure   ____ **6.** Meatus   ____ **9.** Tuberosity

2. Group each of the following bones into one of the four major bone categories. Use *L* for long bone, *S* for short bone, *F* for flat bone, and *I* for irregular bone. Enter the appropriate letter in the space provided.

   ____ **1.** Calcaneus ____ **4.** Humerus   ____ **7.** Radius

   ____ **2.** Frontal   ____ **5.** Mandible   ____ **8.** Sternum

   ____ **3.** Femur   ____ **6.** Metacarpal ____ **9.** Vertebra

3. Using the key choices, characterize the following statements relating to long bones. Enter the appropriate term or letter in the answer blanks.

KEY CHOICES:

A. Diaphysis   **C.** Epiphysis   **E.** Yellow marrow cavity

B. Epiphyseal plate   **D.** Red marrow

_____ **1.** Site of spongy bone in the adult

_____ **2.** Site of compact bone in the adult

_____ **3.** Site of hemopoiesis in the adult

_____ **4.** Scientific name for bone shaft

_____ **5.** Site of fat storage in the adult

_____ **6.** Site of longitudinal growth in a child

4. Complete the following statements concerning bone formation and destruction, using the terms provided in the key. Insert the key letter or corresponding term in the answer blanks.

KEY CHOICES:

A. Atrophy   **C.** Gravity   **E.** Osteoclasts   **G.** Parathyroid hormone

B. Calcitonin   **D.** Osteoblasts   **F.** Osteocytes   **H.** Stress and/or tension

_____ **1.** When blood calcium levels begin to drop below homeostatic levels, **(1)** is released, causing calcium to be released from bones.

_____ **2.** Mature bone cells, called **(2)** , maintain bone in a viable state.

_____ **3.** Disuse such as that caused by paralysis or severe lack of exercise results in muscle and bone **(3)** .

_____ **4.** Large tubercles and/or increased deposit of bony matrix occur at sites of **(4)** .

_____ **5.** Immature, or matrix-depositing, bone cells are referred to as **(5)** .

_____ **6.** **(6)** causes blood calcium to be deposited in bones as calcium salts.

_____ **7.** Bone cells that liquefy bone matrix and release calcium to the blood are called **(7)** .

_____ **8.** Our astronauts must do isometric exercises when in space because bones atrophy under conditions of weightlessness or lack of **(8)** .

5. Five descriptions of bone structure are provided in Column A. First identify the structure by choosing the appropriate term from Column B and placing it or the corresponding letter in the space provided. Then consider Figure 5-1, a diagrammatic view of a cross section of bone. Select different colors for the structures and bone areas in Column B and use them to color the coding circles and corresponding structures in the figure.

Column A

_____ 1. Concentric layers of calcified matrix

_____ 2. Site of osteocytes

_____ 3. Longitudinal canal, carrying blood vessels and nerves

_____ 4. Nonliving, structural part of bone

_____ 5. Minute canals, connecting lacunae

Column B

A. Haversian canal ◯

B. Concentric lamellae ◯

C. Lacunae ◯

D. Canaliculi ◯

E. Matrix ◯

S Jeng

Figure 5-1

6. Figure 5-2 is a longitudinal view of a sagitally sectioned femur. Leave the articular cartilage *white*, but select different colors for the bone regions listed below. Color the coding circles and the corresponding regions on the figure. Complete the illustration by labeling compact bone and spongy bone where indicated by leader lines.

◯ Diaphysis   ◯ Area where red marrow is found

◯ Epiphyseal plate   ◯ Area where yellow marrow is found

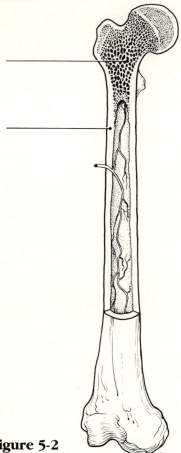

Figure 5-2

Axial Skeleton

Skull

7. Using key choices, identify the bones indicated by the following descriptions. Enter the appropriate term or letter in the answer blanks.

KEY CHOICES:

A. Ethmoid   **E.** Mandible   **I.** Palatines   **L.** Temporals

B. Frontal   **F.** Maxillae   **J.** Parietals   **M.** Vomer

C. Hyoid   **G.** Nasals   **K.** Sphenoid   **N.** Zygomatic

D. Lacrimals   **H.** Occipital

_____ 1. Forehead bone

_____ 2. Cheekbone

_____ 3. Lower jaw

_____ 4. Bridge of nose

_____ 5. Posterior part of hard palate

_____ 6. Much of the lateral and superior cranium

_____ 7. Most posterior part of cranium

_____ 8. Single, irregular, bat-shaped bone, forming part of the cranial floor

_____ 9. Tiny bones, bearing tear ducts

_____ 10. Anterior part of hard palate

_____ 11. Superior and medial nasal conchae formed from its projections

_____ 12. Site of mastoid process

_____ 13. Site of sella turcica

_____ 14. Site of cribriform plate

_____ 15. Site of mental foramen

_____ 16. Site of styloid process

_____ 17. _____ 18. Four bones, containing paranasal sinuses

_____ 19. _____ 20.

_____ 21. Its condyles articulate with the atlas

_____ 22. Foramen magnum contained here

_____ 23. Middle ear found here

_____ 24. Nasal septum

_____ 25. Bears an upward protrusion, the "cock's comb," or crista galli

8. Figure 5-3 (A-C) shows lateral, anterior, and inferior views of the skull. Select different colors for the bones listed below and color the coding circles and corresponding bones in the figure. Complete the figure by labeling the bone markings indicated by leader lines.

◯ Frontal   ◯ Sphenoid   ◯ Zygomatic   ◯ Nasal

◯ Parietal   ◯ Ethmoid   ◯ Palatine   ◯ Lacrimal

◯ Mandible   ◯ Temporal   ◯ Occipital   ◯ Vomer

◯ Maxilla

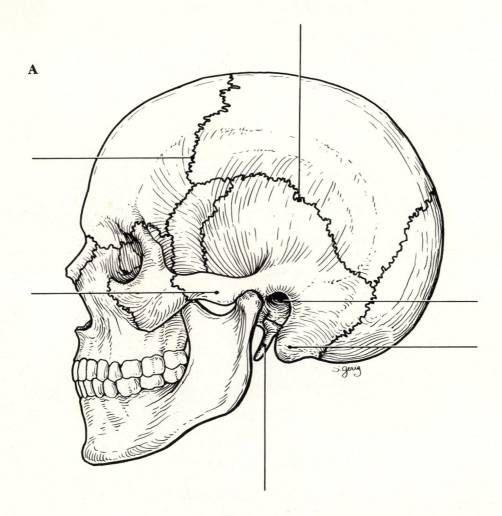

A

Figure 5-3

B

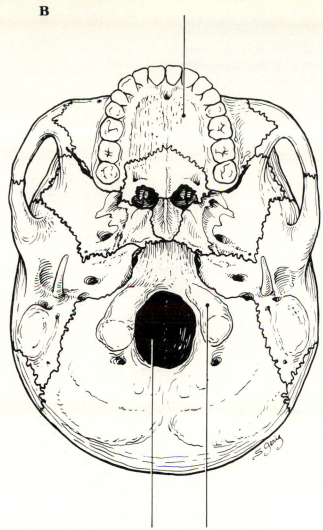

C

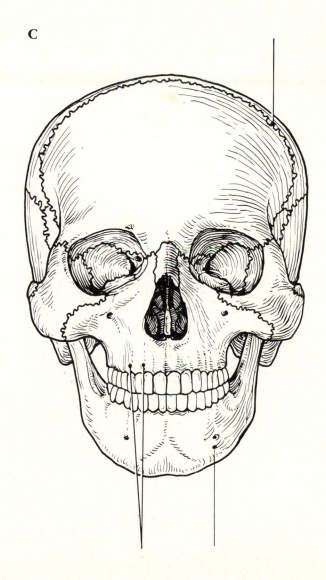

9. An anterior view of the skull, showing the positions of the sinuses, is shown in
   Figure 5-4. First select different colors for each of the sinuses and use them to
   color the coding circles and the corresponding structures on the figure. Then
   briefly answer the following questions concerning the sinuses.

   1. What *are* sinuses? _____

   2. What purpose do they serve in the skull? _____

   _____

   3. Why are they so susceptible to infection? _____

   _____

   ○ Sphenoid sinus   ○ Ethmoid sinuses

   ○ Frontal sinus   ○ Maxillary sinus

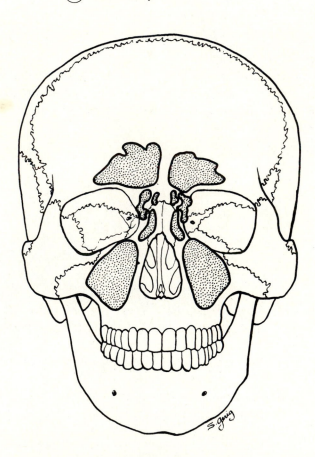

Figure 5-4

Vertebral Column (Spine)

10. Using the key choices, correctly identify the vertebral parts/areas described as follows. Enter the appropriate term(s) or letter(s) in the spaces provided.

KEY CHOICES:

A. Body   **C.** Spinous process   **E.** Transverse process

B. Intervertebral foramina   **D.** Superior articular process   **F.** Vertebral arch

_____ **1.** Structure that encloses the nerve cord

_____ **2.** Weight-bearing portion of the vertebra

_____ **3.** Provide(s) levers for the muscles to pull against

_____ **4.** Provide(s) an articulation point for the ribs

_____ **5.** Openings providing for exit of spinal nerves

11. The following statements provide distinguishing characteristics of the vertebrae composing the vertebral column. Using key choices, identify each described structure or region by inserting the appropriate term(s) or letter(s) in the spaces provided.

KEY CHOICES:

A. Atlas   **D.** Coccyx   **F.** Sacrum

B. Axis   **E.** Lumbar vertebra   **G.** Thoracic vertebra

C. Cervical vertebra—typical

_____ **1.** Type of vertebra(e) containing foramina in the transverse processes, through which the vertebral arteries ascend to reach the brain

_____ **2.** Its dens provides a pivot for rotation of the first cervical vertebra

_____ **3.** Transverse processes have facets for articulation with ribs; spinous process points sharply downward

_____ **4.** Composite bone; articulates with the hipbone laterally

_____ **5.** Massive vertebrae; weight-sustaining

_____ **6.** "Tail-bone"; vestigal fused vertebrae

_____ **7.** Supports the head; allows the rocking motion of the occipital condyles

_____ **8.** Seven components; unfused

_____ **9.** Twelve components; unfused

12. Complete the following statements by inserting your answers in the answer
    blanks.

    _____ **1.**   In describing abnormal curvatures, it could be said that
                                              __(1)__ is an exaggerated thoracic curvature, and in __(2)__
    _____ **2.**   the vertebral column is displaced laterally.

    _____ **3.**   Invertebral discs are made of __(3)__ tissue. The discs provide
                                              __(4)__ to the spinal column.
    _____ **4.**

13. Figure 5-5 (A-D) shows four types of vertebrae. In the spaces provided below
    each vertebra, indicate in which region of the spinal column it would be found.
    In addition, specifically identify Figure 5-5(A).

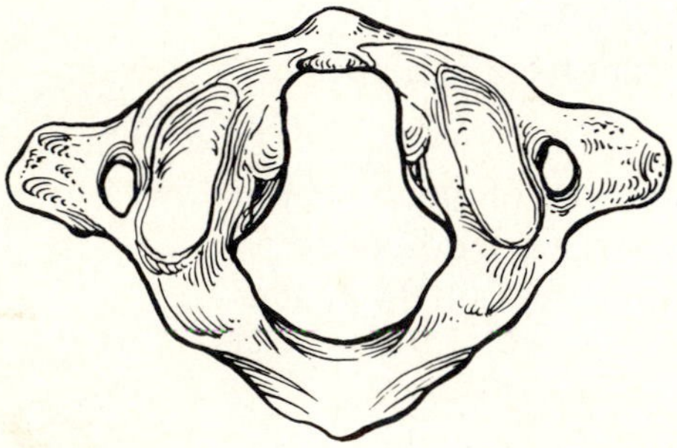

A 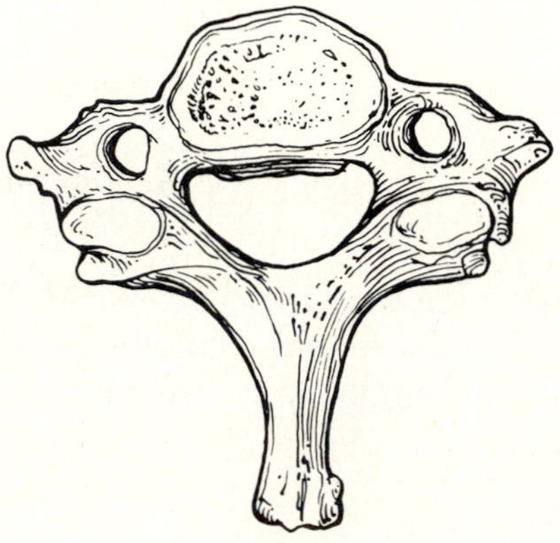 _____   **B** _____

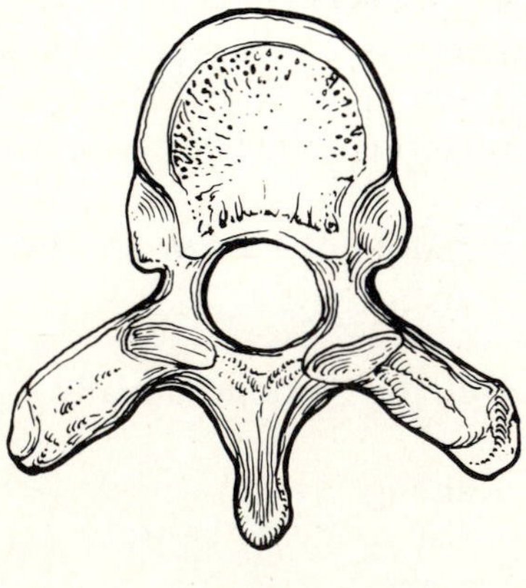

C 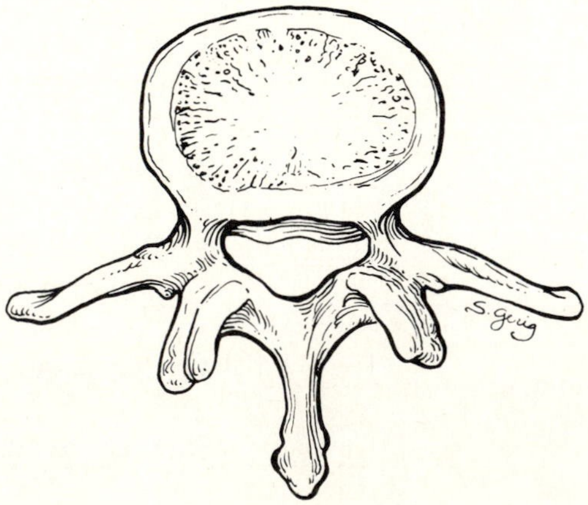 _____   **D** _____

Figure 5-5

14. Figure 5-6 is a lateral view of the vertebral column. Identify each numbered region of the column by listing in the numbered answer blanks the region name first and then the specific vertebrae involved (for example, sacral region, S# to S#). Also identify the modified vertebrae indicated by numbers 6 and 7 in Figure 5-6. Select different colors for each vertebral region and use them to color the coding circles and the corresponding regions.

1. _____ ◯

2. _____ ◯

3. _____ ◯

4. _____ ◯

5. _____ ◯

6. _____ ◯

7. _____ ◯

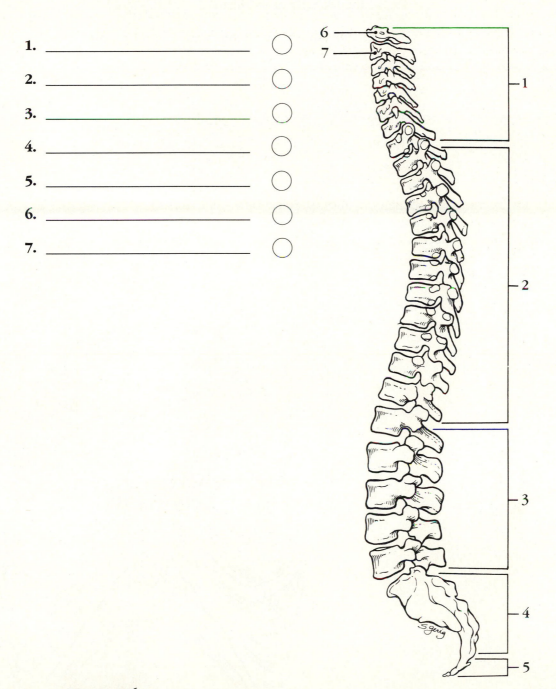

Figure 5-6

Bony Thorax

15. Complete the following statements referring to the bony thorax, by inserting
    your responses in the answer blanks.

    _____ 1.

    _____ 2.

    _____ 3.

    _____ 4.

    _____ 5.

    _____ 6.

    _____ 7.

    _____ 8.

    The organs protected by the thoracic cage include the **(1)**
    and the **(2)** . Ribs 1 through 7 are called **(3)** ribs,
    whereas ribs 8 through 12 are called **(4)** ribs. Ribs 11 and
    12 are also called **(5)** ribs. All ribs articulate posteriorly
    with the **(6)** , and most connect anteriorly to the **(7)** ,
    either directly or indirectly.

    The general shape of the thoracic cage is **(8)** .

16. Figure 5-7 is an anterior view of the bony thorax. Select different colors to
    identify the structures below and color the coding circles and corresponding
    structures. Then label the subdivisions of the sternum indicated by leader lines.

    ○ All true ribs   ○ All false ribs

    ○ Costal cartilages   ○ Sternum

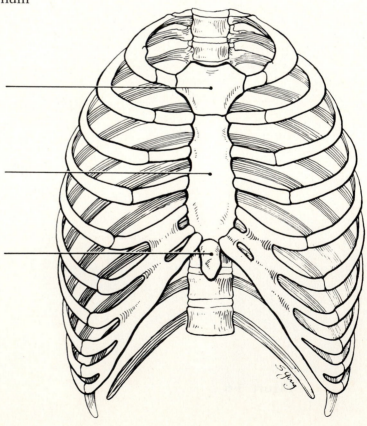

Figure 5-7

Appendicular Skeleton

Several bones forming part of the upper limb and/or shoulder girdle are shown in Figures 5-8 to 5-11. Follow the specific directions for each figure.

17. Identify the bone in Figure 5-8. Insert your answer in the blank below the illustration. Select different colors for each structure listed below and use them to color the coding circles and the corresponding structures in the diagram.

⃝ Spine   ⃝ Glenoid fossa

⃝ Coracoid process

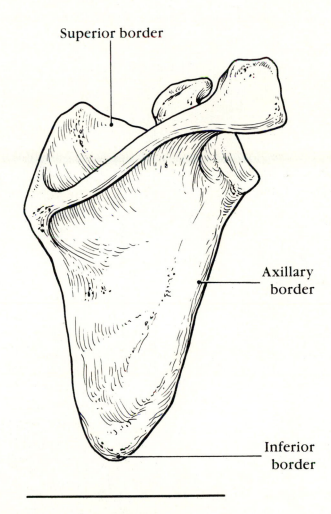

Superior border

Axillary border

Inferior border

Figure 5-8

18. Identify the bone in Figure 5-9. Insert your answer in the answer blank below the illustration. Select different colors for each structure listed and use them to color the coding circles and the corresponding structures in the diagram. Then complete the illustration by inserting a leader line and labeling the deltoid tuberosity.

⃝ Head   ⃝ Capitulum   ⃝ Trochlea   ⃝ Shaft

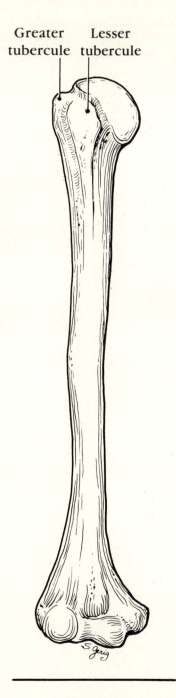

Greater   Lesser
tubercule   tubercule

Figure 5-9

19. Identify the two bones in Figure 5-10 by labeling the leader lines identified as (A) and (B). Color the bones different colors. Using the following terms, complete the illustration by labeling all bone markings provided with leader lines.

⬭ Olecranon process   ⬭ Semilunar notch

⬭ Coronoid process   ⬭ Radial tuberosity

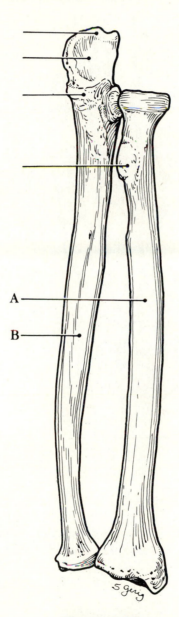

Figure 5-10

20. Figure 5-11 is a diagram of the hand. Select different colors for the following
    structures, and use them to color the coding circles and the corresponding
    structures in the diagram.

    ◯ Carpals   ◯ Phalanges

    ◯ Metacarpals

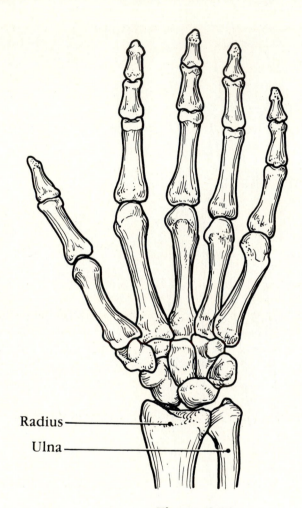

Figure 5-11

21. Compare the pectoral and pelvic girdles by choosing descriptive terms from the
    key choices. Insert the appropriate key letter in the answer blanks.

    KEY CHOICES:

    A. Flexibility   **D.** Shallow socket for limb attachment

    B. Massive   **E.** Deep, secure socket for limb attachment

    C. Lightweight   **F.** Weight-bearing

    Pectoral: _____, _____, _____   Pelvic: _____, _____, _____

22. Using key choices, identify the bone names or markings according to the descriptions that follow. Insert the appropriate term or letter in the answer blanks.

KEY CHOICES:

A. Acromion process	**F.** Coronoid fossa	**K.** Olecranon fossa	**P.** Scapula
B. Capitulum	**G.** Deltoid tuberosity	**L.** Olecranon process	**Q.** Sternum
C. Carpals	**H.** Glenoid fossa	**M.** Phalanges	**R.** Styloid process
D. Clavicle	**I.** Humerus	**N.** Radial tuberosity	**S.** Trochlea
E. Coracoid process	**J.** Metacarpals	**O.** Radius	**T.** Ulna

_____ 1. Raised area on lateral surface of humerus to which deltoid muscle attaches

_____ 2. Arm bone

_____ 3. _____ 4. Bones composing the shoulder girdle

_____ 5. _____ 6. Forearm bones

_____ 7. Point where scapula and clavicle connect

_____ 8. Shoulder girdle bone that has no attachment to the axial skeleton

_____ 9. Shoulder girdle bone that articulates anteriorly with the sternum

_____ 10. Socket in the scapula for the arm bone

_____ 11. Process above the glenoid fossa that permits muscle attachment

_____ 12. Commonly called the collarbone

_____ 13. Distal medial process of the humerus; joins the ulna

_____ 14. Medial bone of the forearm in anatomic position

_____ 15. Rounded knob on the humerus that articulates with the radius

_____ 16. Anterior depression; superior to the trochlea; receives part of the ulna when the forearm is flexed

_____ 17. Forearm bone involved in formation of elbow joint

_____ 18. _____ 19. Bones that articulate with the clavicle

_____ 20. Bones of the wrist

_____ 21. Bones of the fingers

_____ 22. Heads of these bones form the knuckles

23. Using key choices, identify the bone names and markings, according to the descriptions that follow. Insert the appropriate key term(s) or letter(s) in the answer blanks.

KEY CHOICES:

A. Acetabulum	**I.** Ilium	**Q.** Patella
B. Calcaneus	**J.** Ischial tuberosity	**R.** Pubic symphysis
C. Femur	**K.** Ischium	**S.** Pubis
D. Fibula	**L.** Lateral malleolus	**T.** Sacroiliac joint
E. Gluteal tuberosity	**M.** Lesser sciatic notch	**U.** Talus
F. Greater sciatic notch	**N.** Medial malleolus	**V.** Tarsals
G. Greater and lesser trochanters	**O.** Metatarsals	**W.** Tibia
H. Iliac crest	**P.** Obturator foramen	**X.** Tibial tuberosity

_____ **1.** Fuse to form the coxal bone (hip bone)

_____ **2.** Receives the weight of the body when sitting

_____ **3.** Point where the coxal bones join anteriorly

_____ **4.** Upper margin of iliac bones

_____ **5.** Deep socket in the hip bone that receives the head of the thigh bone

_____ **6.** Point where axial skeleton attaches to the pelvic girdle

_____ **7.** Longest bone in body, articulates with the coxal bone

_____ **8.** Lateral bone of the leg

_____ **9.** Medial bone of the leg

_____ **10.** Bones forming the knee joint

_____ **11.** Point where the patellar tendon attaches

_____ **12.** Kneecap

_____ **13.** Shinbone

_____ **14.** Distal process on medial tibial surface

_____ **15.** Process forming the outer ankle

_____ **16.** Heel bone

_____ **17.** Bones of the ankle

_____ **18.** Bones forming the instep of the foot

_____ **19.** Opening in a coxal bone formed by the pubic and ischial rami

_____ **20.** Sites of muscle attachment on the proximal end of the femur

_____ **21.** Tarsal bone that articulates with the tibia

24. Figure 5-12 is a diagram of the articulated pelvis. Identify the bones and bone markings indicated by leader lines on the figure. Select different colors for the structures listed below and use them to color the coding circles and the corresponding structures in the figure. Also, label the dotted line showing the dimensions of the true pelvis and that showing the diameter of the false pelvis. Complete the illustration by labeling the following bone markings: obturator foramen, iliac crest, ischial spine, pubic ramus, and pelvic brim. Last, list three ways in which the female pelvis differs from the male pelvis and insert your answers in the answer blanks.

◯ Coxal bone   ◯ Pubic symphysis

◯ Sacrum   ◯ Acetabulum

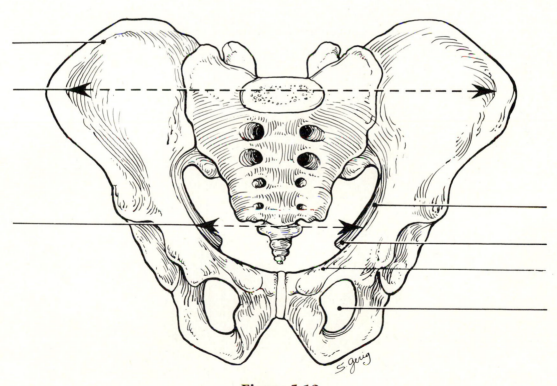

Figure 5-12

1. _____

2. _____

3. _____

25. The bones of the thigh and the leg are shown in Figure 5-13. Identify each and put your answers in the blanks below the diagrams. Select different colors for those structures listed below that are accompanied by a color-coding circle and use them to color in the coding circles and corresponding structures on the diagram. Complete the illustration by inserting the terms indicating bone markings at the ends of the appropriate leader lines in the figure.

◯ Femur   ◯ Tibia   ◯ Fibula

   Head of femur   Anterior crest of tibia   Head of fibula

   Intercondylar eminence   Lesser trochanter   Medial malleolus

   Tibial tuberosity   Greater trochanter   Lateral malleolus

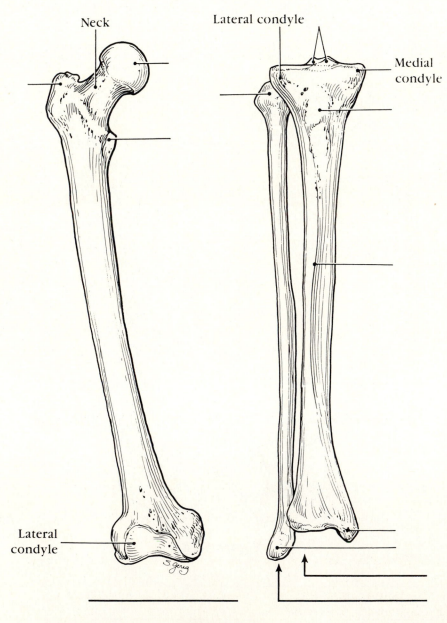

Figure 5-13

26. If the following statements are true, insert *T* in the answer blanks. If any of the statements are false, correct the underlined term(s) by inserting the correct term in the answer blank.

_____ 1. The <u>pectoral</u> girdle is formed by the articulation of the hip bones and the sacrum.

_____ 2. Bones present in both the hand and the foot are <u>carpals</u>.

_____ 3. The tough, fibrous connective tissue covering of a bone is the <u>periosteum</u>.

_____ 4. The point of fusion of the three bones forming a coxal bone is the <u>glenoid fossa</u>.

_____ 5. The large nerve that must be avoided when giving injections into the buttock muscles is the <u>femoral</u> nerve.

_____ 6. The long bones of a fetus are constructed of <u>hyaline</u> cartilage.

_____ 7. Bones that provide the most protection to the abdominal viscera are the <u>ribs</u>.

_____ 8. The largest foramen in the skull is the <u>foramen magnum</u>.

27. Circle the term that does not belong in each of the following groupings.

   1. Tibia   Ulna   Fibula   Femur

   2. Body   Spinous process   Transverse process   Atlas

   3. Hemopoiesis   Red marrow   Yellow marrow   Spongy bone

   4. Manubrium   Body   Xiphoid process   Styloid process

   5. Skull   Rib cage   Vertebral column   Pelvis

   6. Ischium   Scapula   Ilium   Pubis

   7. Mandible   Frontal bone   Temporal bone   Occipital bone

   8. Meatus   Foramen   Condyle   Fissure

   9. Cribriform plate   Ethmoid   Cock's comb   Sphenoid

   10. Canaliculi   Marrow cavity   Volkmann's canals   Haversian canals

28. Figure 5-14 on p. 74 is a diagram of the articulated skeleton. Identify all bones or groups of bones by writing the correct labels at the end of the leader lines. Then, select two different colors for the bones of the axial and appendicular skeletons and use them to color in the coding circles and corresponding structures in the diagram.

⭕ Axial skeleton   ⭕ Appendicular skeleton

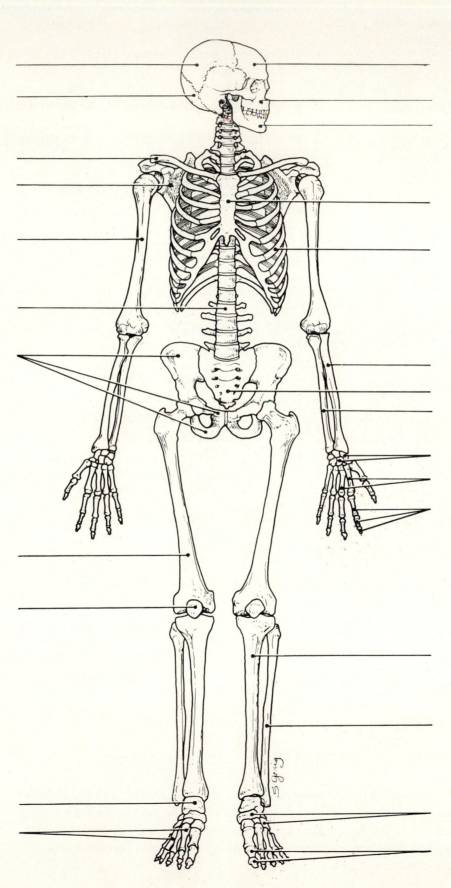

Figure 5-14

Bone Fractures

29. Using the key choices, identify the fracture types and treatments described below. Enter the appropriate key letter in each answer blank.

KEY CHOICES:

A. Closed reduction   C. Compound fracture   E. Open reduction

B. Compression fracture   D. Greenstick fracture   F. Simple fracture

_____ 1. Bone is broken cleanly; the ends do not penetrate the skin

_____ 2. Nonsurgical realignment of broken bone ends and splinting of bone

_____ 3. A break common in children; bone splinters, but break is incomplete

_____ 4. A fracture in which the bone is crushed; common in the vertebral column

_____ 5. A fracture in which the bone ends penetrate through the skin surface

_____ 6. Surgical realignment of broken bone ends

30. For each of the following statements about bone breakage and the repair process that is true, insert *T* in the answer blank. For false statements, correct the underlined terms by inserting the correct term in the answer blank.

_____ 1. A <u>hematoma</u> usually forms at a fracture site.

_____ 2. Deprived of nutrition, <u>osteocytes</u> at the fracture site die.

_____ 3. Non-bony debris at the fracture site is removed by <u>osteoclasts</u>.

_____ 4. Growth of a new capillary supply into the region produces <u>granulation tissue</u>.

_____ 5. Osteoblasts from the <u>medullary cavity</u> migrate to the fracture site.

_____ 6. The <u>fibrocartilage callus</u> is the first repair mass to splint the broken bone.

_____ 7. The bony callus is composed of <u>compact</u> bone.

Joints, or Articulations

31. Use the key choices to identify the joint types described below. Insert the appropriate key term or corresponding letter in the answer blanks.

   KEY CHOICES:

   A. Amphiarthrosis   **B.** Diarthrosis   **C.** Synarthrosis

   _____ **1.** Allows slight degree of movement

   _____ **2.** Essentially immovable

   _____ **3.** Characterized by cartilage connecting the bony portions

   _____ **4.** All have a fibrous capsule lined with a synovial membrane surrounding a joint cavity

   _____ **5.** Freely movable

   _____ **6.** Bone regions united by fibrous connective tissue

   _____ **7.** Joint between skull bones

   _____ **8.** Joint between the axis and atlas

   _____ **9.** Hip joint

   _____ **10.** Intervertebral joints

   _____ **11.** Elbow joint

   _____ **12.** Intercarpal joints

   _____ **13.** Pubic symphysis

   _____ **14.** Interphalangeal joints

   _____ **15.** Often reinforced by ligaments

32. Complete the following statements by inserting your answer(s) in the numbered answer blanks.

   _____ **1.** Diarthrotic joints often contain fluid-filled cushions called __(1)__ .

   _____ **2.** Displacement of a bone out of its socket is __(2)__ .

   _____ **3.** Ball and socket joints allow the greatest degree of __(3)__ .

   _____ **4.** __(4)__ joints provide the most protection to underlying structures.

33. Figure 5-15 shows the structure of a typical diarthrotic joint. Select different colors to identify each of the following areas and use them to color the coding circles and the corresponding structures on the figure.

   ◯ Articular cartilage of bone ends   ◯ Synovial membrane

   ◯ Fibrous capsule   ◯ Joint cavity

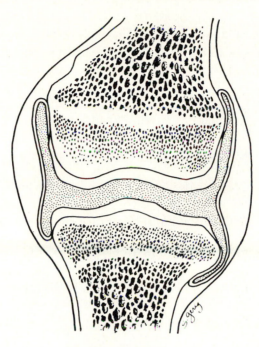

Figure 5-15

Homeostatic Imbalances of Bones and Joints

34. For each of the following statements that is true, enter *T* in the answer blank. For each false statement, correct the underlined words by writing the correct words in the answer blank.

   _____ 1. In a sprain, the ligaments reinforcing a joint are excessively stretched or torn.

   _____ 2. Age-related erosion of articular cartilages and formation of painful bony spurs are characteristic of gouty arthritis.

   _____ 3. Chronic arthritis usually results from bacterial invasion.

   _____ 4. Healing of a partially torn ligament is slow because its hundreds of fibrous strands are poorly aligned.

   _____ 5. Rheumatoid arthritis is an autoimmune disease.

   _____ 6. High levels of uric acid in the blood may lead to rheumatoid arthritis.

_____   **7.** A "soft" bone condition in children, usually due to a lack of calcium or vitamin D in the diet, is called <u>osteomyelitis</u>.

_____   **8.** Atrophy and thinning of bone owing to hormonal changes or inactivity (generally in the elderly) is called <u>osteoporosis</u>.

Developmental Aspects of the Skeletal System

35. Using the key choices, identify the body systems that relate to bone tissue viability. Enter the appropriate key terms or letters in the answer blanks.

KEY CHOICES:

A. Endocrine   **C.** Muscular   **E.** Reproductive

B. Integumentary   **D.** Nervous   **F.** Urinary

_____   **1.** Conveys the sense of pain in bone and joints

_____   **2.** Activates vitamin D for proper calcium usage

_____   **3.** Regulates uptake and release of calcium by bones

_____   **4.** Increases bone strength and viability by pulling action

_____   **5.** Influences skeleton proportions and adolescent growth of long bones

_____   **6.** Provides vitamin D for proper calcium absorption

36. Complete the following statements concerning fetal and infant skeletal development. Insert the missing words in the answer blanks.

_____   **1.**   "Soft spots," or membranous joints called **(1)** in the fetal skull, allow the skull to be **(2)** slightly during birth
_____   **2.**   passage. They also allow for continued brain **(3)** during the later months of fetal development and early infancy.
_____   **3.**   Eventually these soft spots are replaced by immovable joints called **(4)** .
_____   **4.**

_____   **5.**   The two spinal curvatures present at birth are the **(5)** and **(6)** curvatures. Because they are present at birth, they are called **(7)** curvatures. The secondary curvatures
_____   **6.**   develop as the baby matures. The **(8)** curvature develops as the baby begins to lift his or her head. The **(9)**
_____   **7.**   curvature develops when the baby begins to walk or assume the upright posture.

_____   **8.**

_____   **9.**

The Incredible Journey:
A Visualization Exercise
for the Skeletal System

. . . stalagmite- and stalactite-like structures that surround you. . . . Since the texture is so full of holes. . .

37. Where necessary, complete statements by inserting the missing words in the answer blanks.

_____ 1.

_____ 2.

_____ 3.

_____ 4.

_____ 5.

_____ 6.

_____ 7.

_____ 8.

_____ 9.

_____ 10.

_____ 11.

_____ 12.

For this journey you are miniaturized and injected into the interior of the largest bone of your host's body, the __(1)__. Once inside this bone, you look around and find yourself examining the stalagmite- and stalactite-like structures that surround you. Although you feel as if you are in an underground cavern, you know that it has to be bone. Since the texture is so full of holes, it obviously is __(2)__ bone. Although the arrangement of these bony spars seems to be haphazard, as if someone randomly dropped straws, they are precisely arranged to resist points of __(3)__. All about you is frantic, hurried activity. Cells are dividing rapidly, nuclei are being ejected, and disklike cells are appearing. You decide that these disklike cells are __(4)__, and that this is the __(5)__ cavity. As you explore further, strolling along the edge of the cavity, you spot many tunnels leading into the solid bony area on which you are walking. Walking into one of these drainpipe-like openings, you notice that it contains a glistening white ropelike structure (a __(6)__, no doubt), and blood vessels running the length of the tube. You eventually come to a point in the channel where the horizontal passageway joins with a vertical passage that runs with the longitudinal axis of the bone. This is obviously a __(7)__ canal. Since you would like to see how nutrients are brought into __(8)__ bone, you decide to follow this channel. Reasoning that there is no way you can possibly scale the slick walls of the channel, you leap and grab onto a white cord hanging down its length. Since it is easier to slide down than to try to climb up the cord, you begin to lower yourself, hand over hand. During your descent, you notice small openings in the wall, which are barely large enough for you to wriggle through. You conclude that these are the __(9)__ that connect all the __(10)__ to the nutrient supply in the central canal. You decide to investigate one of these tiny openings and begin to swing on your cord, trying to get a foothold on one of the openings. After managing to anchor yourself and squeezing into an opening, you use a flashlight to illuminate the passageway in front of you. You are startled by a giant cell with many dark nuclei. It appears to be plastered around the entire lumen directly ahead of you. As you watch this cell the bony material beneath it, the __(11)__, begins to liquefy. The cell apparently is a bone-digesting cell, or __(12)__, and since you are unsure whether or not its enzymes can also liquefy you, you slither backwards hurriedly and begin your trek back to your retrieval site.

Muscular System

Muscles, the specialized tissues that facilitate body movement, make up about 40% of body weight. The bulk of body muscle is the voluntary type, called skeletal muscle because it is attached to the bony skeleton. Voluntary muscle contributes to body contours and shape, and this muscle type composes the organ system called the muscular system. These muscles allow you to smile, frown, run, swim, shake hands, grasp things, and to otherwise manipulate your environment. The balance of body muscle is smooth and cardiac muscles, which form the bulk of the walls of hollow organs and the heart. Smooth and cardiac muscles are involved in the transport of materials within the body.

Study activities in this chapter deal with microscopic and gross structure of muscle, identification of voluntary muscles, body movements, and important understandings of muscle physiology.

Structure of Muscle Tissues

1. Nine characteristics of muscle tissue are listed below. Identify the muscle tissue type described by choosing the correct response(s) from the key choices. Enter the appropriate term(s) or letter(s) of the key choice in the answer blank.

   KEY CHOICES:

   A. Cardiac   **B.** Smooth   **C.** Skeletal

   _____A, B_____   **1.** Involuntary

   _____A, C_____   **2.** Banded appearance

   _____B_____   **3.** Longitudinally and circularly arranged layers

_____C_____   **4.** Dense connective tissue packaging

_____A_____   **5.** Figure-8 packaging of the cells

_____A_____   **6.** Coordinated activity to act as a pump

_____C_____   **7.** Moves bones and the facial skin

_____C_____   **8.** Referred to as the muscular system

_____G_____   **9.** Voluntary

2. First, identify the structures described in Column A by matching them with the terms provided in Column B. Enter the correct term or letter in the answer blanks. Then, select a different color for each of the terms in Column B provided with a color-coding circle and color them in on Figure 6-1.

Column A

_____G_____   **1.** Connective tissue, ensheathing a bundle of muscle cells

_____B_____   **2.** Another term for deep fascia

_____I_____   **3.** Contractile unit of muscle

_____D_____   **4.** A muscle cell

_____A_____   **5.** Thin, reticular connective tissue investing each muscle cell

_____H_____   **6.** Cell membrane of the muscle cell

_____F_____   **7.** A long, filamentous organelle found within muscle cells; has a banded appearance

_____E_____   **8.** Actin- or myosin-containing structure

_____K_____   **9.** Cordlike extension of connective tissue beyond the muscle, serving to attach it to the bone

Column B

A. Endomysium ◯

B. Epimysium ◯

C. Fascicle

D. Myofiber ◯ _cell_

E. Myofilament _actin myosin_

F. Myofibril ◯ _contractile tissue_

G. Perimysium ◯

H. Sarcolemma _membrane_

I. Sarcomere _contracts_

J. Sarcoplasm

K. Tendon ◯

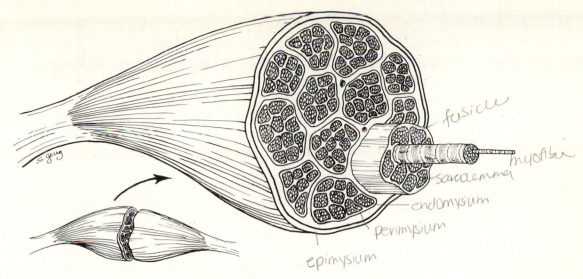

Figure 6-1

3. Figure 6-2 is a diagrammatic representation of a small portion of a muscle cell (bracket indicates the portion enlarged). First, select different colors for the structures listed below. Use them to color the coding circles and corresponding structures on Figure 6-2. When you have finished, bracket an A band and an I band.

○ Myosin   ○ Actin filaments

○ Z line

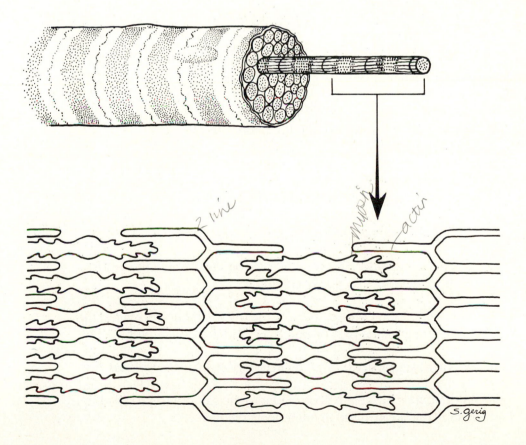

Figure 6-2

Muscle Activity

4. Complete the following statements relating to the myoneural junction. Insert the correct answers in the numbered answer blanks.

_____nervo muscular_____ **1.**

_____motor unit_____ **2.**

_____ **3.**

_____synaptic cleft_____ **4.**

_____acetocholine_____ **5.**

_____impulse_____ **6.**

_____ **7.**

The junction between a motor neuron's axon and the muscle cell membrane is called a myoneural junction, or a __(1)__ junction.

A motor neuron and all of the skeletal muscle cells it stimulates is called a __(2)__ . The axon of each motor neuron has numerous endings called __(3)__ .

The actual gap between an axonal ending and the muscle cell is called a __(4)__ . Within the axonal endings are many small vesicles, containing a neurotransmitter substance called __(5)__ .

When the __(6)__ reaches the ends of the axon, the neurotransmitter is released, and it diffuses to the muscle cell membrane to combine with receptors there.

The combination of the neurotransmitters with the muscle membrane receptors causes the membrane to become permeable to sodium, resulting in the influx of sodium ions and __(7)__ of the membrane. Then contraction of the muscle cell occurs.

5. Number the following statements in their proper sequence to describe the contraction mechanism in a skeletal muscle cell. The first step has already been identified as number 1.

___1___ **1.** Acetylcholine is released into the neuromuscular junction by the axonal terminal.

_____ **2.** The action potential, carried deep into the cell, causes the sarcoplasmic reticulum to release calcium ions.

_____ **3.** The muscle cell relaxes and lengthens.

_____ **4.** Acetylcholine diffuses across the neuromuscular junction and binds to receptors on the sarcolemma.

_____ **5.** The calcium ion concentration at the myofilaments increases; the myofilaments slide past one another, and the cell shortens.

_____ **6.** Depolarization occurs, and the action potential is generated.

_____ **7.** As calcium is actively reabsorbed into the sarcoplasmic reticulum, its concentration at the myofilaments decreases.

6. The following incomplete statements refer to a muscle cell in the resting, or polarized, state just before stimulation. Complete each statement by choosing the correct response from the key choices and entering the appropriate letter in the answer blanks.

KEY CHOICES:

A. Na^+ diffuses out of the cell

B. K^+ diffuses out of the cell

C. Na^+ diffuses into the cell

D. K^+ diffuses into the cell

E. Inside the cell

F. Outside the cell

G. Relative ionic concentrations on the two sides of the membrane during rest

H. Electrical conditions

I. Activation of the sodium-potassium pump, which moves K^+ into the cell and Na^+ out of the cell

J. Activation of the sodium-potassium pump, which moves Na^+ into the cell and K^+ out of the cell

_____ 1.

_____ 2.

_____ 3.

_____ 4.

_____ 5.

_____ 6.

_____ 7.

There is a greater concentration of Na^+ __(1)__ , and there is a greater concentration of K^+ __(2)__ . When the stimulus is delivered, the permeability of the membrane is changed, and __(3)__ , initiating the depolarization of the membrane. Almost as soon as the depolarization wave begins, a repolarization wave follows it across the membrane. This occurs as __(4)__ . Repolarization restores the __(5)__ of the resting cell membrane. The __(6)__ is (are) reestablished by __(7)__ .

7. Which of the following occur within a muscle cell during oxygen debt? Place a check (√) by the correct choices.

_____ **1.** Decreased ATP

_____ **2.** Increased ATP

_____ **3.** Increased lactic acid

_____ **4.** Decreased oxygen

_____ **5.** Increased oxygen

_____ **6.** Decreased carbon dioxide

_____ **7.** Increased carbon dioxide

_____ **8.** Increased glucose

8. Briefly describe when you can tell you are repaying the oxygen debt.

_____

_____

9. Complete the following statements by choosing the correct response from the
    key choices and entering the appropriate letter or term in the answer blanks.

    KEY CHOICES:

A. Fatigue	**E.** Isometric contraction	**I.** Many motor units
B. Isotonic contraction	**F.** Whole muscle	**J.** Repolarization
C. Muscle cell	**G.** Tetanus	**K.** Depolarization
D. Muscle tone	**H.** Few motor units	

    _____G_____   1. _____ is a continuous contraction that shows no evidence of
                           relaxation.

    _____B_____   2. A(n) _____ is a contraction in which the muscle shortens and work
                           is done.

    _____E_____   3. To accomplish a strong contraction, _____ are stimulated at a rapid
                           rate.

    _____C_____   4. The "all-or-none" law applies to skeletal muscle function at
                           the _____ level.

    _____J_____   5. The absolute refractory period is the time when a muscle cell
                           cannot be stimulated because _____ is occurring.

    _____H_____   6. When a weak but smooth muscle contraction is desired, _____ are
                           stimulated at a rapid rate.

    _____A_____   7. When a muscle is being stimulated but is not able to respond due
                           to "oxygen debt," the condition is called _____ .

    _____E_____   8. A(n) _____ is a contraction in which the muscle does not shorten
                           but tension in the muscle keeps increasing.

Body Movements and Naming Skeletal Muscles

10. Complete the following statements. Insert your answers in the answer blanks.

    _____insertion_____   1.   The movable attachment of a muscle is called its __(1)__ ,
                                 and the stationary attachment is called the __(2)__ .

    _____origin_____   2.

                                 Winding up for a pitch (as in baseball) can properly be
    _____   3.   called __(3)__ .

    _____adduct_____   4.   To keep your seat when riding a horse, the tendency is
                                 to __(4)__ your thighs.

_____flex_____ 5.

_____ext_____ 6.

_____ext_____ 7.

_____flexed_____ 8.

_____flexion_____ 9.

_____ 10.

_____ 11.

_____ 12.

_____ 13.

_____ 14.

_____ 15.

_____ 16.

In running, the action at the hip joint is __(5)__ in reference to the leg moving forward and __(6)__ in reference to the leg in the posterior position. When kicking a football, the action at the knee is __(7)__ .

In climbing stairs, the hip and knee of the forward leg are both __(8)__ .

You have just touched your chin to your chest; this is __(9)__ of the neck.

Using a screwdriver with a straight arm requires __(10)__ of the arm.

Consider all the movements of which the arm is capable. One often used for strengthening all the upper arm and shoulder muscles is __(11)__ .

Moving the head to signify "no" is __(12)__ .

Standing on your toes as in ballet requires __(13)__ of the foot.

Action that moves the distal end of the radius across the ulna is __(14)__ .

Raising the arms laterally away from the body is called __(15)__ of the arms.

Walking on your heels is __(16)__ .

11. The terms provided in the key are often applied when muscles are discussed according to the manner in which they interact with other muscles. Select the key terms that apply to the following definitions and insert the correct letter or term in the answer blanks.

KEY CHOICES:

A. Antagonist   **B.** Fixator   **C.** Prime mover   **D.** Synergist

_____C_____ **1.** Agonist

_____B_____ **2.** Postural muscles for the most part

_____D_____ **3.** Stabilizes a joint so that the prime mover can act at more distal joints

_____D_____ **4.** Performs the same movement as the prime mover

_____A_____ **5.** Reverses and/or opposes the action of a prime mover

_____B_____ **6.** Immobilizes the origin of a prime mover

12. Several criteria are applied to the naming of muscles. These are provided in the key choices. Identify which ones pertain to the muscles listed here and enter the correct letter(s) in the answer blank.

    KEY CHOICES:

    A. Action of the muscle

    B. Shape of the muscle

    C. Location of the muscle's origin and/or insertion

    D. Number of origins

    E. Location of muscle relative to a bone or body region

    F. Direction in which the muscle fibers run relative to some imaginary line

    G. Relative size of the muscle

    _____ 1. Gluteus maximus

    _G_ 2. Adductor magnus

    _D_ 3. Biceps femoris

    _F_ 4. Abdominis transversus

    _____ 5. Extensor carpi ulnaris

    _B_ 6. Trapezius

    _____ 7. Rectus femoris

    _____ 8. External oblique

Gross Anatomy of the Skeletal Muscles

Muscles of the Head

13. Name the major muscles described here. Select a different color for each muscle listed and color in the coding circle and corresponding muscles on Figure 6-3.

    ○ ___zygomaticus major___   Used in smiling

    ○ ___buccinator___   Used to suck in your cheeks

    ○ ___orbicularis occulus___   Used in winking

    ○ ___frontalis___   Used to form the horizontal frown crease on the forehead or to raise your eyebrows

    ○ ___orbicularis oris___   The "kissing" muscle

    ○ ___masseter___   Prime mover of jaw closure

    ○ ___temporalus___   Synergist muscle for jaw closure

    ○ ___sternocledu mastoid___   Prime mover of head flexion; a two-headed muscle

Zygomatic bone

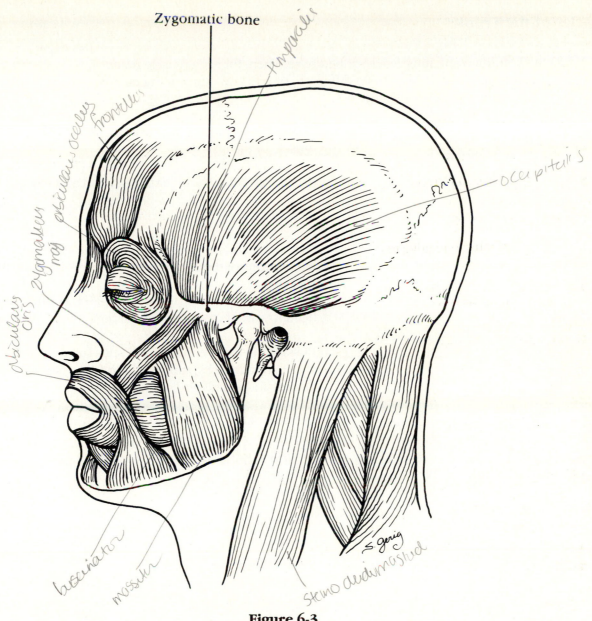

temporalis

frontalis

obicularis oculus

occipitalis

zygomaticus maj

obicularis oris

buccinator

masseter

sterno cleidomastid

S. Gerig

Figure 6-3

Muscles of the Trunk

14. Name the anterior trunk muscles described here. Then, for each muscle name
that has a color coding circle, select a different color to color the coding circle
and corresponding muscle on Figure 6-4.

○ _____rectus abdominus_____   A major spine flexor; the name means "straight
muscle of the abdomen"

○ _____pectralis major_____   Prime mover for shoulder flexion and adduction

○ _____deltoid_____   Prime mover for shoulder abduction

○ _____extumal oblique_____   Part of the abdominal girdle; forms the external
lateral walls of the abdomen

○ _____Sternocleido mastoid_____   Acting alone, each muscle of this pair turns the
head toward the opposite shoulder

_____internal oblique_____   Besides the two abdominal muscles (pairs) named
above, two muscle pairs that help form the
_____transverse abdominis_____   natural abdominal girdle

_____   Deep muscles of the thorax that promote the
inspiratory phase of breathing

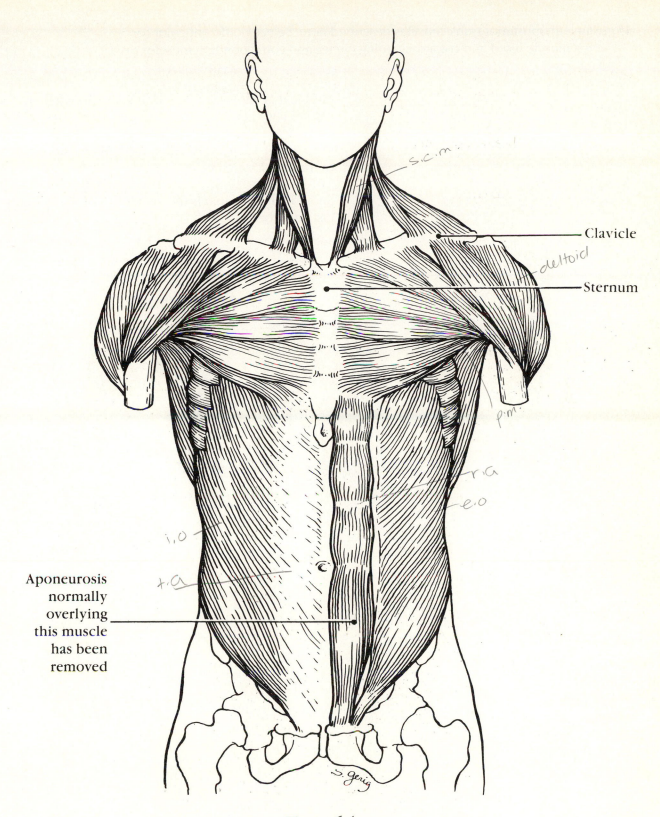

Clavicle

Sternum

Aponeurosis
normally
overlying
this muscle
has been
removed

Figure 6-4

15. Name the posterior trunk muscles described here. Select a different color for
each muscle listed and color the coding circles and corresponding muscles on
Figure 6-5.

○ _____trapezius_____   Muscle that allows you to shrug your shoulders or
extend your head

○ _____latissimus dorsi_____   Muscle that adducts shoulder and causes extension
of the shoulder joint

○ _____deltoid_____   Shoulder muscle that is the antagonist of the muscle
just described

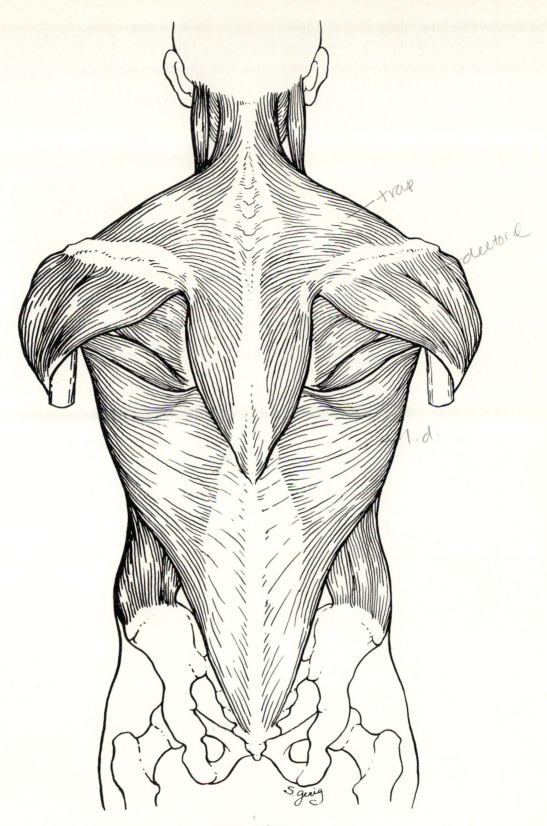

Figure 6-5

Muscles of the Hip, Thigh, and Leg

16. Name the muscles described here. Select a different color for each muscle
    provided with a color-coding circle, and use it to color the coding circles and
    corresponding muscles on Figure 6-6. Complete the illustration by labeling
    those muscles provided with leader lines.

_____   Hip flexor, deep in pelvis

◯ _____   Used to extend the hip when climbing
stairs; forms buttock

◯ _____   "Toe dancer's" muscle

◯ _____   Inverts and dorsiflexes the foot

◯ _____   Allows you to draw your legs to the midline
of your body, as when standing at attention

◯ _____   Muscle group that extends the knee

◯ _____   Muscle group that extends the thigh and
flexes the knee

◯ _____   Smaller hip muscle commonly used as
an injection site

_____   Muscle group of the lateral leg; plantar flex
and evert the foot

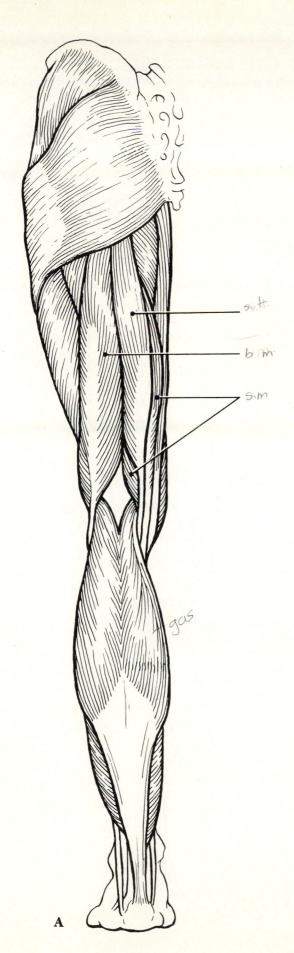

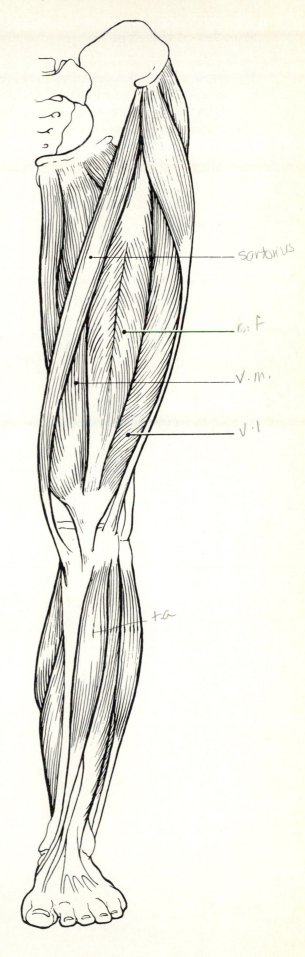

A **B**

Figure 6-6

Muscles of the Arm

17. Name the muscles described here. Then select different colors for each muscle provided with a color-coding circle and use them to color in the coding circles and corresponding muscles on Figure 6-7.

Wrist flexor that follows the ulna

◯ _____

Muscle that extends the fingers

◯ _____

Muscle that flexes the fingers

_____

Muscle that allows you to bend (flex) the elbow

◯ _____

Muscle that extends the elbow

◯ _____

Powerful shoulder abductor, used to raise the arm overhead

◯ _____

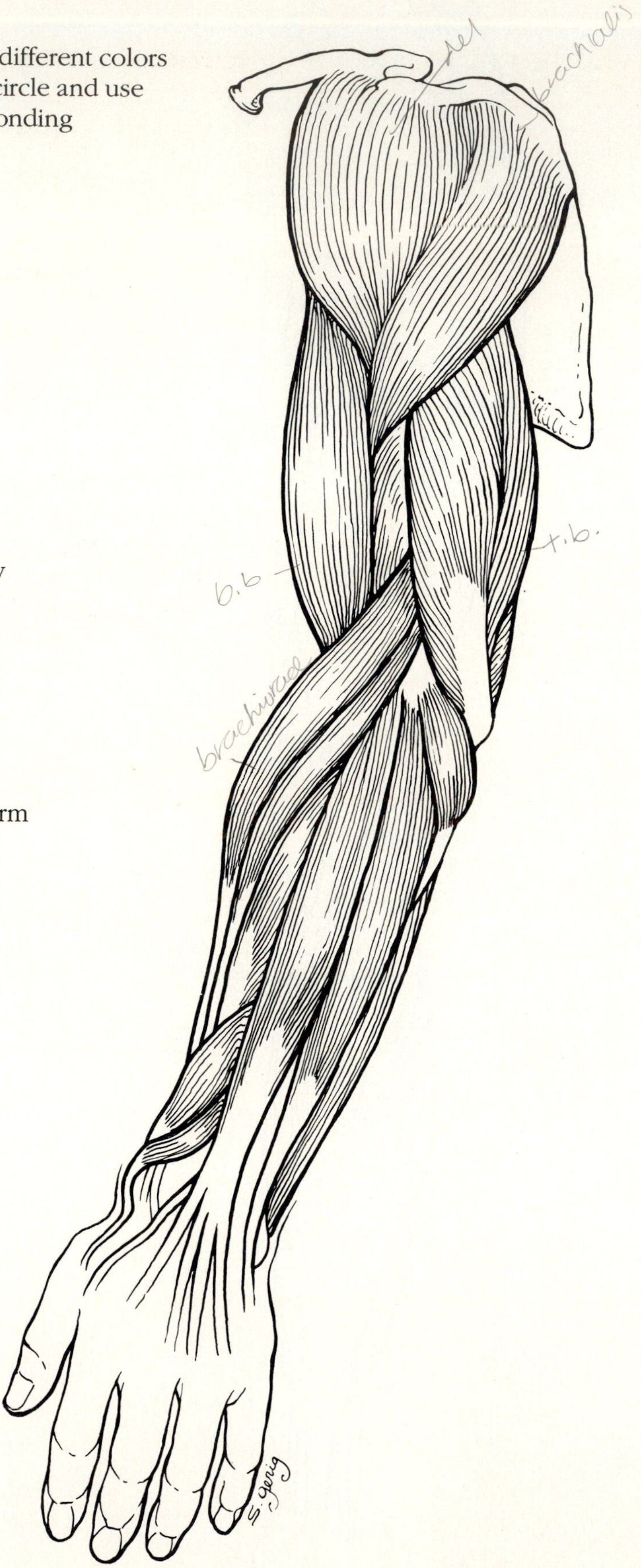

Figure 6-7

General Body Muscle Review

18. Complete the following statements describing muscles. Insert the correct answers in the answer blanks.

_____ 1.

_____ 2.

_____ 3.

_____ 4.

_____ 5.

_____ 6.

_____ 7.

_____ 8.

_____ 9.

_____ 10.

Three muscles— **(1)** , **(2)** , and **(3)** —are commonly used for intramuscular injections.

The insertion tendon of the **(4)** group contains a large sesamoid bone, the patella.

The triceps surae insert in common into the **(5)** tendon.

The bulk of the tissue of a muscle tends to lie **(6)** to the part of the body it causes to move.

The extrinsic muscles of the hand originate on the **(7)** .

Most flexor muscles are located on the **(8)** aspect of the body; most extensors are located **(9)** . An exception to this generalization is the extensor–flexor musculature of the **(10)** .

19. Identify the numbered muscles in Figure 6-8 by placing the numbers in the blanks next to the following muscle names. Then select a different color for each muscle provided with a color-coding circle and color the coding circle and corresponding muscle in Figure 6-8.

           ____ **1.** Orbicularis oris

◯ ____ **2.** Pectoralis major

◯ ____ **3.** External oblique

◯ ____ **4.** Sternocleidomastoid

◯ ____ **5.** Biceps brachii

◯ ____ **6.** Deltoid

◯ ____ **7.** Vastus lateralis

◯ ____ **8.** Frontalis

◯ ____ **9.** Rectus femoris

◯ ____ **10.** Sartorius

◯ ____ **11.** Gracilis

◯ ____ **12.** Adductor group

           ____ **13.** Peroneus longus

           ____ **14.** Temporalis

◯ ____ **15.** Orbicularis oculi

◯ ____ **16.** Zygomaticus

           ____ **17.** Masseter

           ____ **18.** Vastus medialis

           ____ **19.** Tibialis anterior

◯ ____ **20.** Transversus abdominus

           ____ **21.** Tensor fascia lata

◯ ____ **22.** Rectus abdominis

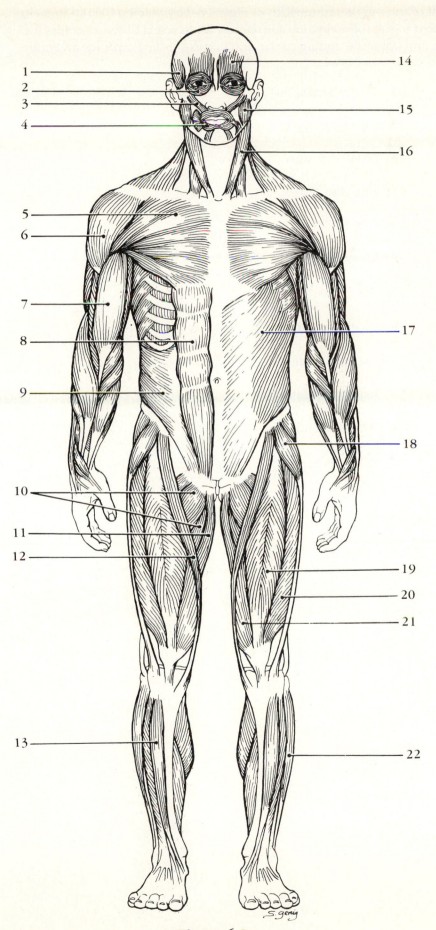

Figure 6-8

20. Identify each of the numbered muscles in Figure 6-9 by placing the numbers in the blanks next to the following muscle names. Then select different colors for each muscle and color the coding circles and corresponding muscles on Figure 6-9.

○   _____   **1.** Gluteus maximus

○   _____   **2.** Adductor muscle

○   _____   **3.** Gastrocnemius

○   _____   **4.** Latissimus dorsi

○   _____   **5.** Deltoid

○   _____   **6.** Semitendinosus

○   _____   **7.** Trapezius

○   _____   **8.** Biceps femoris

○   _____   **9.** Triceps brachii

○   _____   **10.** External oblique

○   _____   **11.** Gluteus medius

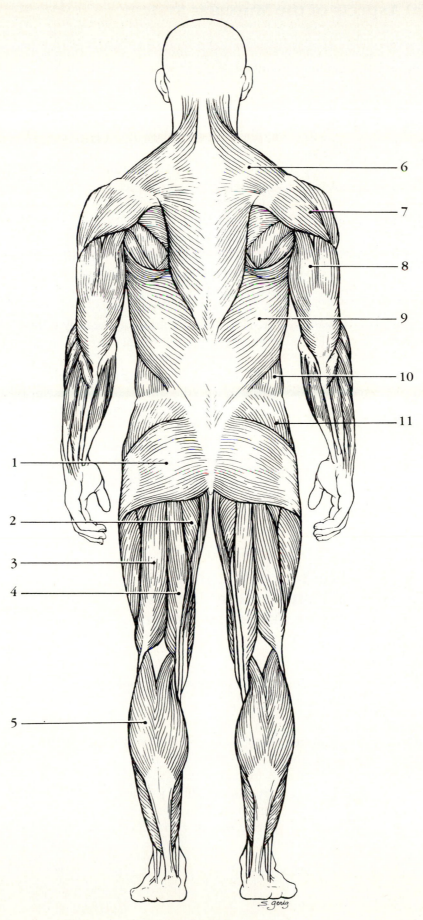

Figure 6-9

Developmental Aspects of the Muscular System

21. Complete the following statements concerning the embryonic development of muscles and their functioning throughout life. Insert your answers in the answer blanks.

1. _____

2. _____

3. _____

4. _____

5. _____

6. _____

7. _____

8. _____

9. _____

10. _____

11. _____

12. _____

The first movement of the baby detected by the mother-to-be is called the __(1)__ .

An important congenital muscular disease that results in the degeneration of the skeletal muscles by young adulthood is called __(2)__ .

A baby's control over muscles progresses in a __(3)__ direction as well as a __(4)__ direction. In addition, __(5)__ muscular control (that is, waving of the arms) occurs before __(6)__ control (pincer grasp) does.

Muscles will ordinarily stay healthy if they are __(7)__ regularly; without normal stimulation they __(8)__ .

__(9)__ is a disease of the muscles, which results from some problem with the stimulation of muscles by acetylcholine. The muscles become progressively weaker in this disease.

With age, our skeletal muscles decrease in mass; this leads to a decrease in body __(10)__ and in muscle __(11)__ . Muscle tissue that is lost is replaced by noncontractile __(12)__ tissue.

The Incredible Journey: A Visualization Exercise for the Muscular System

As you straddle this structure, you wonder what is happening.

22. Where necessary, complete statements by inserting the missing words in the numbered spaces.

1. _____

2. _____

On this incredible journey, you will be miniaturized and enter a skeletal muscle cell to observe the events that occur during muscle contraction. You prepare yourself by donning

_____ 3.

_____ 4.

_____ 5.

_____ 6.

_____ 7.

_____ 8.

_____ 9.

_____ 10.

a wet suit and charging your ion detector. Then you climb into a syringe to prepare for injection. Your journey will begin when you see the gleaming connective tissue covering, the __(1)__ of a single muscle cell. Once injected, you monitor your descent through the epidermis and subcutaneous tissue. When you reach the muscle cell surface, you see that it is punctuated with pits at relatively regular intervals. Looking into the darkness and off in the distance, you can see a leash of fibers ending close to a number of muscle cells. Since all of these fibers must be from the same motor neuron, this functional unit is obviously a __(2)__. You approach the fiber ending on your muscle cell and scrutinize the __(3)__ junction there. As you examine the junction, minute fluid droplets leave the nerve ending and attach to doughnut-shaped receptors on the muscle cell membrane. This substance released by the nerve ending must be __(4)__. Then, as a glow falls over the landscape, your ion detector indicates ions are disappearing from the muscle cell exterior and entering the muscle pits. The needle drops from high to low as the __(5)__ ions enter the pits from the watery fluid outside. You should have expected this, since these ions must enter to depolarize the muscle cells and start the __(6)__.

Next you begin to explore one of the surface pits. As the muscle jerks into action, you topple deep into the pit. Sparkling electricity lights up the wall on all sides. You grasp for a handhold. Finally successful, you pull yourself laterally into the interior of the muscle cell and walk carefully along what seems to be a log. Then, once again, you notice an eerie glow as your ion detector reports that __(7)__ ions are entering the cytoplasm rapidly. The "log" you are walking on "comes to life" and begins to slide briskly in one direction. Unable to keep your balance, you fall. As you straddle this structure, you wonder what is happening. On all sides, cylindrical structures—such as the one you are astride—are moving past other similar but larger structures. Suddenly you remember, these are the myofilaments, __(8)__ and __(9)__, that slide past one another during muscle contraction.

Seconds later, the forward movement ends, and you begin to journey smoothly in the opposite direction. The ion detector now indicates low __(10)__ ion levels. Since you cannot ascend the smooth walls of one of the entry pits, you climb from one myofilament to another to reach the underside of the sarcolemma. Then you travel laterally to enter a pit close to the surface and climb out onto the cell surface. Your journey is completed, and you prepare to leave your host once again.

Nervous System

The nervous system is the master coordinating system of the body. Every thought, action, and sensation reflects its activity. Because of its complexity, the anatomic structures of the nervous system are considered in terms of two principal divisions—the central nervous system (CNS) and the peripheral nervous system (PNS). The CNS, consisting of the brain and spinal cord, interprets incoming sensory information and issues instructions based on past experience. The PNS, consisting of cranial and spinal nerves and ganglia, serves as the communication line between the CNS and the body's muscles, glands, and sensory receptors. The nervous system is also divided functionally in terms of motor activities into the somatic and autonomic divisions. It is important, however, to recognize that these classifications are made for the sake of convenience and that the nervous system acts in an integrated manner both structurally and functionally.

Student activities provided in this chapter consider neuron anatomy and physiology, identification of the various structures of the central and peripheral nervous system, reflex and sensory physiology, and a summary of autonomic nervous system anatomy and physiology. Because every body system is controlled, at least in part, by the nervous system, understanding of these areas is extremely important to the understanding of the overall functioning of the body.

1. List the three major functions of the nervous system.

1. _____

   _____

2. _____

   _____

3. _____

   _____

Organization of the Nervous System

2. Choose the key responses that best correspond to the descriptions provided in the following statements. Insert the appropriate letter or term in the answer blanks.

KEY CHOICES:

A. Autonomic nervous system   C. Peripheral nervous system (PNS)

B. Central nervous system (CNS)   D. Somatic nervous system

_____ **1.** Nervous system subdivision that is composed of the brain and spinal cord.

_____ **2.** Subdivision of the PNS that controls voluntary activities such as the activation of skeletal muscles.

_____ **3.** Nervous system subdivision that is composed of the cranial and spinal nerves and ganglia.

_____ **4.** Subdivision of the PNS that regulates the activity of the heart and smooth muscle and of glands; it is also called the involuntary nervous system.

_____ **5.** A major subdivision of the nervous system that interprets incoming information and issues orders.

_____ **6.** A major subdivision of the nervous system that serves as the communication lines, linking all parts of the body to the CNS.

Nervous Tissue—Structure and Function

3. This exercise emphasizes the difference between neurons and neuroglia. Indicate which cell type is identified by the following descriptions by inserting the appropriate letter or term in the answer blanks.

KEY CHOICES:

A. Neurons   B. Neuroglia

_____ **1.** Supports, insulates, and protects cells

_____ **2.** Demonstrate irritability and conductivity, and thus transmit electrical messages from one area of the body to another area

_____ **3.** Release neurotransmitters

_____ **4.** Are amitotic

_____ **5.** Able to divide; therefore are responsible for most brain neoplasms

4. Match the anatomic terms given in Column B with the appropriate descriptions of function provided in Column A. Place the correct term or letter response in the answer blanks.

	Column A	Column B
_____	1. Releases neurotransmitters	A. Axon
_____	2. Conducts the impulse toward the nerve cell body	B. Axonal terminal
_____	3. Increases the speed of impulse transmission	C. Dendrite
		D. Myelin sheath
_____	4. Location of the nucleus	E. Cell body
_____	5. Conducts impulses away from the nerve cell body	

5. Certain activities or sensations are provided in the following list. Using key choices, select the specific receptor type that would be activated by the activity or sensation described. Insert the correct term(s) or letter response(s) in the answer blanks. Note that more than one receptor type may be activated in some cases.

KEY CHOICES:

A. Bare nerve endings (pain)   E. Muscle spindle

B. Golgi tendon organ   F. Pacinian corpuscle

C. Krause's end bulb   G. Ruffini's corpuscle

D. Meissner's corpuscle

Activity or Sensation	Receptor Type
Walking on hot pavement	1. (Identify two) _____ and _____
Feeling a pinch	2. (Identify two) _____ and _____
Leaning on a shovel	3. _____
Muscle sensations when rowing a boat	4. (Identify two) _____ and _____
Feeling a caress	5. _____

6. List in order the *minimum* elements in a reflex arc from the stimulus to the activity of the effector. Place your responses in the answer blanks.

    1. Stimulus   **4.** _____

    2. _____   **5.** Effector organ

    3. _____

7. Using key choices, select the terms identified in the following descriptions by inserting the appropriate letter or term in the spaces provided.

    KEY CHOICES:

A. Afferent neuron	**F.** Neuroglia	**K.** Proprioceptors
B. Association neuron	**G.** Neurotransmitters	**L.** Schwann cells
C. Cutaneous sense organs	**H.** Nerve	**M.** Synapse
D. Efferent neuron	**I.** Nodes of Ranvier	**N.** Stimuli
E. Ganglion	**J.** Nuclei	**O.** Tract

    _____ **1.** Sensory receptors found in the skin, which are specialized to detect temperature, pressure changes, and pain

    _____ **2.** Specialized cells that myelinate the fibers of neurons found in the PNS

    _____ **3.** Junction or point of close contact between neurons

    _____ **4.** Bundle of nerve processes inside the CNS

    _____ **5.** Neuron, serving as part of the conduction pathway between sensory and motor neurons

    _____ **6.** Gaps in a myelin sheath

    _____ **7.** Collection of nerve cell bodies found outside the CNS

    _____ **8.** Neuron that conducts impulses away from the CNS to muscles and glands

    _____ **9.** Sensory receptors found in muscle and tendons that detect their degree of stretch

    _____ **10.** Changes, occurring within or outside the body, that affect nervous system functioning

_____ **11.** Neuron that conducts impulses toward the CNS from the body periphery

_____ **12.** Chemicals released by neurons that stimulate other neurons, muscles, or glands

8. Figure 7-1 is a diagram of a neuron. First, label the parts indicated on the illustration by leader lines. Then choose different colors for each of the structures listed below and use them to color in the coding circles and corresponding structures in the illustration. Finally, draw arrows on the figure to indicate the direction of impulse transmission along the neuron's membrane.

   ◯ Axon   ◯ Cell body

   ◯ Dendrites   ◯ Myelin sheath

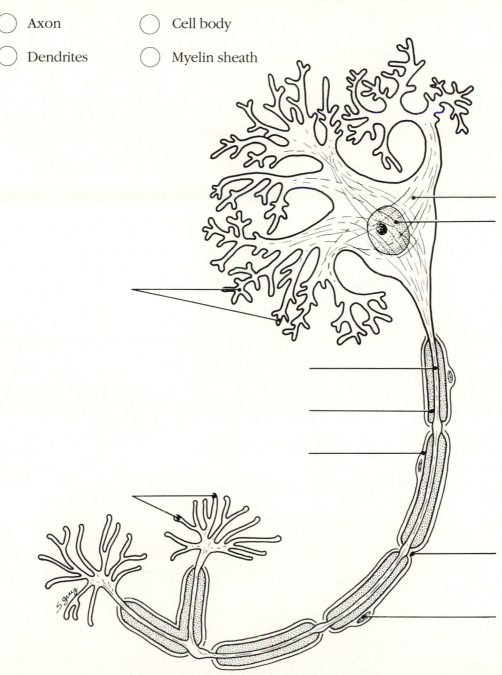

Figure 7-1

9. Using the key choices, identify the terms defined in the following statements.
   Place the correct term or letter response in the answer blanks.

   KEY CHOICES:

   A. Action potential   **D.** Potassium ions   **G.** Sodium ions

   B. Depolarization   **E.** Refractory period   **H.** Sodium–potassium pump

   C. Polarized   **F.** Repolarization

   _____ **1.** Period of repolarization of the neuron during which it cannot respond to a second stimulus

   _____ **2.** State in which the resting potential is reversed as sodium ions rush into the neuron

   _____ **3.** Electrical condition of the plasma membrane of a resting neuron

   _____ **4.** Period during which potassium ions diffuse out of the neuron

   _____ **5.** Transmission of the depolarization wave along the neuron's membrane

   _____ **6.** The chief positive intracellular ion in a resting neuron

   _____ **7.** Process by which ATP is used to move sodium ions out of the cell and potassium ions back into the cell; completely restores the resting conditions of the neuron

10. Using the key choices, identify the types of reflexes involved in each of the following situations.

    KEY CHOICES:

    A. Somatic reflex(es)   **B.** Autonomic reflex(es)

    _____ **1.** Patellar (knee-jerk) reflex

    _____ **2.** Pupillary light reflex

    _____ **3.** Effectors are skeletal muscles

    _____ **4.** Effectors are smooth muscle and glands

    _____ **5.** Flexor reflex

    _____ **6.** Regulation of blood pressure

    _____ **7.** Salivary reflex

11. Refer to Figure 7-2, showing a reflex arc, as you complete the following exercise. First, briefly answer the following questions by inserting your responses in the spaces provided.

   1. What is the stimulus? _____

   2. What tissue is the effector? _____

   3. How many synapses occur in this reflex arc? _____

   Next, select different colors for each of the following structures and use them to color in the coding circles and corresponding structures in the diagram. Finally, draw arrows on the figure indicating the direction of impulse transmission through this reflex pathway.

   ◯ Receptor region   ◯ Association neuron

   ◯ Afferent neuron   ◯ Efferent neuron

   ◯ Effector

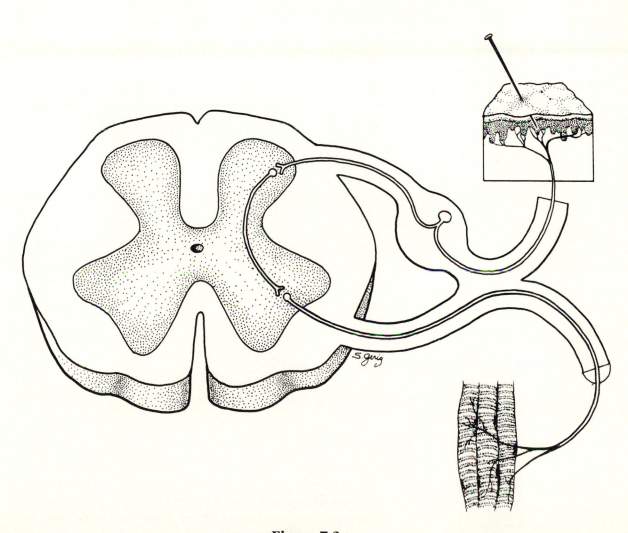

Figure 7-2

12. Circle the term that does not belong in each of the following groupings.

   1. Astrocytes   Neurons   Oligodendrocytes   Microglia

   2. K$^+$ enters the cell   K$^+$ leaves the cell   Repolarization   Refractory period

   3. Dendrite   Axon   End plates   Vesicles containing neurotransmitter

   4. Nodes of Ranvier   Myelin sheath   Unmyelinated   Saltatory conduction

   5. Predictable response   Voluntary act   Involuntary act   Reflex

   6. Oligodendrocytes   Schwann cells   Myelin   Microglia

   7. Cutaneous receptors   Free dendritic endings   Stretch   Pain and touch

   8. Cell interior   High Na$^+$ Low Na$^+$ High K$^+$

Central Nervous System

Brain

13. Complete the following statements by inserting your answers in the answer blanks.

   _____ **1.**

   _____ **2.**

   _____ **3.**

   _____ **4.**

   _____ **5.**

   The largest part of the human brain is the (paired) __(1)__ . The other major subdivisions of the brain are the __(2)__ and the __(3)__ . The cavities found in the brain are called __(4)__ . They contain __(5)__ .

14. Circle the following structures that are definitely *not* part of the brain stem.

   Cerebral hemispheres   Midbrain   Medulla

   Pons   Cerebellum   Diencephalon

15. Complete the following statements by inserting your answers in the answer blanks.

   _____ **1.**

   _____ **2.**

   _____ **3.**

   _____ **4.**

   _____ **5.**

   A __(1)__ is an elevated ridge of cerebral cortex tissue. The convolutions seen in the cerebrum are important because they increase the __(2)__ . Gray matter is composed of __(3)__ . White matter is composed of __(4)__ , which provide for communication between different parts of the brain as well as with lower CNS centers. The lentiform nucleus, the caudate, and other nuclei are collectively called the __(5)__ .

16. Figure 7-3 is a diagram of the right lateral view of the human brain. First, match the letters on the diagram with the following list of terms and insert the appropriate letters in the answer blanks. Then, select different colors for each of the areas of the brain provided with a color-coding circle and use them to color in the coding circles and corresponding structures in the diagram. If an identified area is part of a lobe, use the color you selected for the lobe but use *stripes* for that area.

_____ 1. ◯ Frontal lobe   _____ 7.   Lateral fissure

_____ 2. ◯ Parietal lobe   _____ 8.   Central fissure

_____ 3. ◯ Temporal lobe   _____ 9. ◯ Cerebellum

_____ 4. ◯ Precentral gyrus   _____ 10. ◯ Medulla

_____ 5.   Parieto-occipital fissure.   _____ 11. ◯ Occipital lobe

_____ 6. ◯ Postcentral gyrus   _____ 12. ◯ Pons

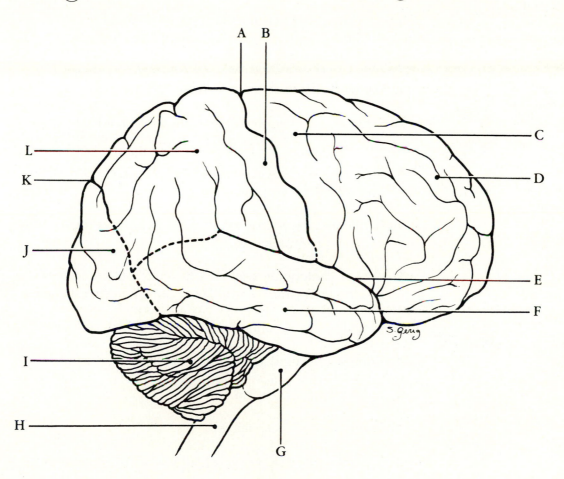

Figure 7-3

17. Figure 7-4 is a diagram of the sagittal view of the human brain. First, match the letters on the diagram with the following list of terms and insert the appropriate letter in the answer blanks. Then, color the brain-stem areas blue and the areas where cerebrospinal fluid is found yellow.

_____ **1.** Cerebellum

_____ **2.** Cerebral aqueduct

_____ **3.** Cerebral hemisphere

_____ **4.** Cerebral peduncle

_____ **5.** Choroid plexus

_____ **6.** Corpus callosum

_____ **7.** Fourth ventricle

_____ **8.** Hypothalamus

_____ **9.** Medulla oblongata

_____ **10.** Optic chiasma

_____ **11.** Pineal body

_____ **12.** Pituitary gland

_____ **13.** Pons

_____ **14.** Thalamus

18. Referring to the brain areas listed in Exercise 17, match the appropriate brain structures with the following descriptions. Insert the correct terms in the answer blanks.

_____ **1.** Site of regulation of water balance and body temperature

_____ **2.** Reflex centers involved in regulating respiratory rhythm in conjunction with lower brainstem centers

_____ **3.** Responsible for the regulation of posture and coordination of skeletal muscle movements

_____ **4.** Important relay station for afferent fibers, traveling to the sensory cortex for interpretation

_____ **5.** Contains autonomic centers, which regulate blood pressure and respiratory rhythm, as well as coughing and sneezing centers

_____ **6.** Large fiber tract, connecting the cerebral hemispheres

_____ **7.** Connects the third and fourth ventricles

_____ **8.** Encloses the third ventricle

_____ **9.** Forms the cerebrospinal fluid

_____ **10.** Midbrain area that is largely fiber tracts; bulges anteriorly

_____ **11.** Part of the limbic system; contains centers for many drives (rage, pleasure, hunger, sex, etc.)

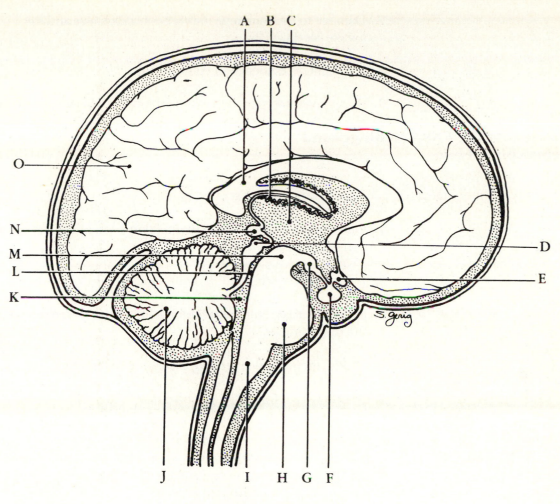

Figure 7-4

19. If the following statements are true, insert *T* in the answer blanks. If any of the statements are false, correct the underlined term by inserting the correct term in the answer blank.

_____ **1.** The primary sensory area of the cerebral hemisphere(s) is found in the underlined precentral gyrus.

_____ **2.** Cortical areas involved in audition are found in the occipital lobe.

_____ **3.** The primary motor area in the temporal lobe is involved in the initiation of voluntary movements.

_____ **4.** The specialized motor speech area is located at the base of the precentral gyrus in an area called Wernicke's area.

_____ **5.** The right cerebral hemisphere receives sensory input from the right side of the body.

_____ **6.** The pyramidal tract is the major descending voluntary motor tract.

_____ 7. Damage to the <u>thalamus</u> impairs consciousness and the awake/sleep cycles.

_____ 8. A <u>flat</u> EEG is evidence of clinical death.

_____ 9. Beta waves are recorded when an individual is awake and <u>relaxed</u>.

Protection of the CNS—Meninges and Cerebrospinal Fluid

20. Identify the meningeal (or associated) structures described here.

_____ 1. Outermost covering of the brain, composed of tough fibrous connective tissue

_____ 2. Innermost covering of the brain; delicate and vascular

_____ 3. Structures that return cerebrospinal fluid to the venous blood in the dural sinuses

_____ 4. Middle meningeal layer; like a cobweb in structure

_____ 5. Its outer layer forms the periosteum of the skull

21. Figure 7-5 shows a frontal view of the meninges of the brain at the level of the superior sagittal (dural) sinus. First, label the *arachnoid villi* on the figure. Then, select different colors for each of the following structures and use them to color the coding circles and corresponding structures in the diagram.

◯ Dura mater   ◯ Pia mater

◯ Arachnoid membrane   ◯ Subarachnoid space

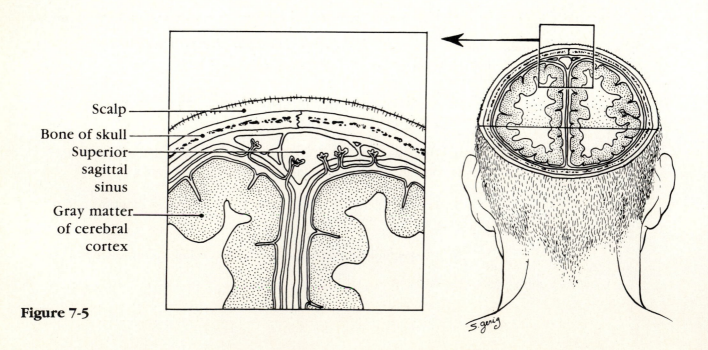

Scalp

Bone of skull

Superior sagittal sinus

Gray matter of cerebral cortex

Figure 7-5

22. Complete the following statements by inserting your answers in the answer blanks.

_____ 1.

_____ 2.

_____ 3.

_____ 4.

_____ 5.

_____ 6.

_____ 7.

Cerebrospinal fluid is formed by capillary knots called __(1)__ , which hang into the __(2)__ of the brain. Ordinarily cerebrospinal fluid flows from the lateral ventricles to the third ventricle and then through the __(3)__ to the fourth ventricle. Some of the fluid continues down the __(4)__ of the spinal cord, but most of it circulates into the __(5)__ by passing through three tiny openings in the walls of the __(6)__ . As a rule, cerebrospinal fluid is formed and drained back into the venous blood at the same rate. If its drainage is blocked, a condition called __(7)__ occurs, which results in increased pressure on the brain.

23. Application of Knowledge. You have been given all of the information needed to identify the brain regions involved in the following situations. See how well your nervous system has integrated this information, and name the brain region (or condition) most likely to be involved in each situation. Place your responses in the answer blanks.

1. Following a train accident, a man with an obvious head injury was observed stumbling about the scene. An inability to walk properly and a loss of balance were quite obvious. What brain region was injured?

_____

2. An elderly woman is admitted to the hospital to have a gallbladder operation. While she is being cared for, the nurse notices that she has trouble initiating movement and has a strange "pill-rolling" tremor of her hands. What cerebral area is most likely involved?

_____

3. A child is brought to the hospital with a high temperature. The doctor states that the child's meninges are inflamed. What name is given to this condition?

_____

4. A young woman is brought into the emergency room with extremely dilated pupils. Her friends state that she has overdosed on cocaine. What cranial nerve is stimulated by the drug?

_____

5. A young man has just received serious burns, resulting from standing with his back too close to a bonfire. He is muttering that he never felt the pain. Otherwise, he would have smothered the flames by rolling on the ground. What part of his CNS might be malfunctional?

_____

6. An elderly gentleman has just suffered a stroke. He is able to understand verbal and written language, but when he tries to respond, his words are garbled. What cortical region has been damaged by the stroke?

_____

7. A 12-year-old boy suddenly falls to the ground, having an epileptic seizure. He is rushed to the emergency room of the local hospital for medication. His follow-up care includes a scan of his brain waves to try to determine the area of the lesion. What is this procedure called?

_____

Spinal Cord

24. Complete the following statements by inserting your responses in the answer blanks.

_____ **1.**

_____ **2.**

_____ **3.**

_____ **4.**

_____ **5.**

_____ **6.**

_____ **7.**

_____ **8.**

_____ **9.**

The spinal cord extends from the __(1)__ of the skull to the __(2)__ region of the vertebral column. The meninges, which cover the spinal cord, extend more inferiorly to form a sac from which cerebrospinal fluid can be withdrawn without damage to the spinal cord. This procedure is called a __(3)__ . __(4)__ pairs of spinal nerves arise from the cord. Of these, __(5)__ pairs are cervical nerves, __(6)__ pairs are thoracic nerves, __(7)__ pairs are lumbar nerves, and __(8)__ pairs are sacral nerves. The taillike collection of spinal nerves at the inferior end of the spinal cord is called the __(9)__ .

25. Using key choices, select the appropriate terms to respond to the following descriptions referring to spinal cord anatomy. Place the correct term or letter in the answer blanks.

KEY CHOICES:

A. Afferent (sensory)   **C.** Both afferent and efferent

B. Efferent (motor)   **D.** Association neurons (interneurons)

_____ **1.** Neuron type found in the dorsal horn

_____ **2.** Neuron type found in the ventral horn

_____ **3.** Neuron type in dorsal root ganglion

_____ **4.** Fiber type in the ventral root

_____ **5.** Fiber type in the dorsal root

_____ **6.** Fiber type in spinal nerve

_____ **7.** Fiber type in white matter of the cord

26. Figure 7-6 is a cross-sectional view of the spinal cord. First, select different colors to identify the following structures and use them to color the coding circles and corresponding structures in the figure.

○ Pia mater ○ Dura mater

○ Arachnoid

Then, identify the areas listed in the key choices by inserting the correct choices/letter next to the appropriate leader line on the figure.

KEY CHOICES:

A. Central canal

B. Columns of white matter

C. Dorsal horn

D. Dorsal root

E. Dorsal root ganglion

F. Spinal nerve

G. Ventral horn

H. Ventral root

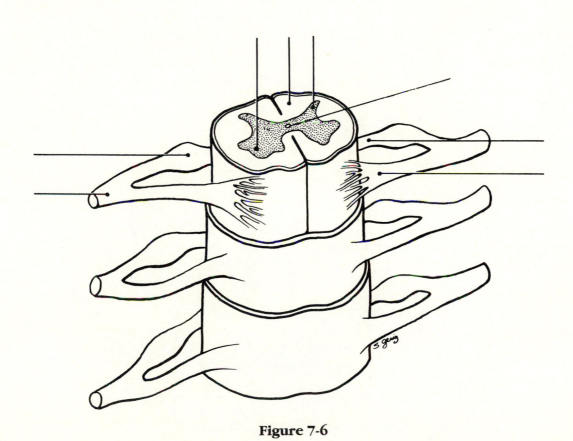

Figure 7-6

27. Using key choices, indicate what would happen if the following structures were damaged or transected. Place the correct letter in the answer blanks.

KEY CHOICES:

A. Loss of motor function   **C.** Loss of both motor and sensory function

B. Loss of sensory function

_____ **1.** Dorsal root of a spinal nerve

_____ **2.** Ventral root of a spinal nerve

_____ **3.** Anterior ramus of a spinal nerve

Peripheral Nervous System

Structure of a Nerve

28. Figure 7-7 is a diagrammatic view of a nerve wrapped in its connective tissue coverings. Select different colors to identify the following structures and use them to color the coding circles and corresponding structures in the figure. Then, label each of the sheaths indicated by leader lines on the figure.

  ◯   Endoneurium   ◯   Perineurium   ◯   Epineurium

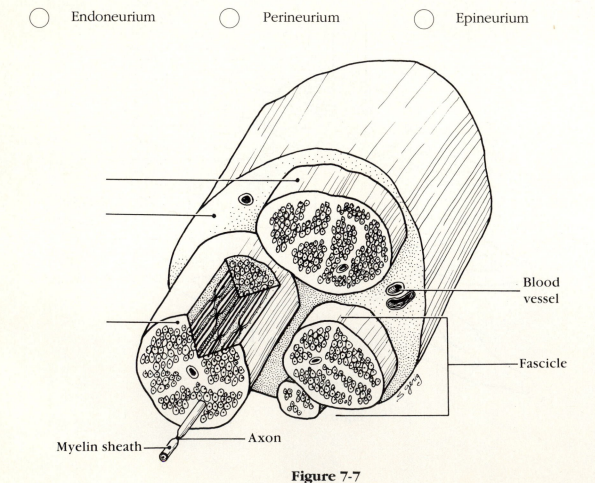

Blood vessel

Fascicle

Myelin sheath —— Axon

Figure 7-7

29. Complete the following statements by inserting your responses in the answer blanks.

_____ 1.

_____ 2.

_____ 3.

Another name for a bundle of nerve fibers is __(1)__ . Nerves carrying both sensory and motor fibers are called __(2)__ nerves, whereas those carrying just sensory fibers are referred to as sensory, or __(3)__ , nerves.

Cranial Nerves

30. The 12 pairs of cranial nerves are indicated by leader lines in Figure 7-8. First, label each by name and Roman numeral on the figure and then color each nerve with a different color.

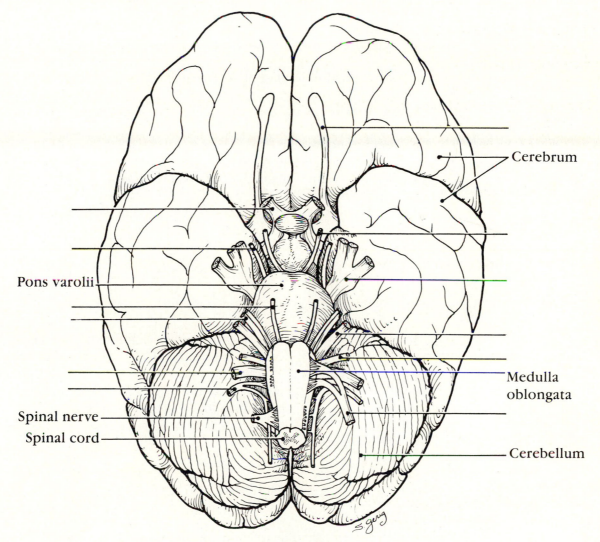

Cerebrum

Pons varolii

Medulla oblongata

Spinal nerve

Spinal cord

Cerebellum

Figure 7-8

31. Provide the name and number of the cranial nerves involved in each of the following activities, sensations, or disorders. Insert your response in the answer blanks.

_____ **1.** Shrugging the shoulders

_____ **2.** Smelling a flower

_____ **3.** Raising the eyelids and focusing the lens of the eye for accommodation; constriction of the eye pupils

_____ **4.** Slows the heart; increases the mobility of the digestive tract

_____ **5.** Involved in smiling

_____ **6.** Involved in chewing food

_____ **7.** Listening to music; seasickness

_____ **8.** Secretion of saliva; tasting well-seasoned food

_____ **9.** Involved in "rolling" the eyes (three nerves—provide numbers only)

_____ **10.** Feeling a toothache

_____ **11.** Reading "Tennis" magazine or this study guide

_____ **12.** Purely sensory (three nerves—provide numbers only)

Spinal Nerves and Nerve Plexuses

32. Complete the following statements by inserting your responses in the answer blanks.

_____ **1.**

_____ **2.**

_____ **3.**

_____ **4.**

The ventral rami of spinal nerves C_1 through T_1 and L_1 through S_4 take part in forming __(1)__ , which serve the __(2)__ of the body. The ventral rami of T_1 through T_{12} run between the ribs to serve the __(3)__ . The posterior rami of the spinal nerves serve the __(4)__ .

33. Name the major nerves that serve the following body areas. Insert your responses in the answer blanks.

_____ **1.** Head, neck, shoulders (name plexus only)

_____ **2.** Diaphragm

_____ **3.** Posterior thigh

_____ **4.** Leg and foot (name two)

_____ **5.** Most anterior forearm muscles

_____ **6.** Arm muscles

_____ **7.** Abdominal wall (name plexus only)

_____ **8.** Anterior thigh

_____ **9.** Medial side of the hand

Autonomic Nervous System

34. The following table indicates a number of conditions. Use a check (✓) to show which division of the autonomic nervous system is involved in each condition.

Condition	Sympathetic	Parasympathetic
1. Postganglionic neurons secrete norepinephrine; adrenergic fibers		
2. Postganglionic neurons secrete acetylcholine; cholinergic fibers		
3. Long preganglionic axon, short postganglionic axon		
4. Short preganglionic axon, long postganglionic axon		
5. Arises from cranial and sacral nerves		
6. Arises from spinal nerves T_1 to L_3		
7. Normally in control		
8. Fight-or-flight system		
9. Has more specific control		
10. Causes a dry mouth, dilates bronchioles		
11. Constricts eye pupils, decreases heart rate		

35. You are alone in your home late in the evening, and you hear an unfamiliar sound in your backyard. In the spaces provided, list four physiologic events promoted by the sympathetic nervous system that would help you to cope with this rather frightening situation.

1. _____

2. _____

3. _____

4. _____

36. After surgery, patients are often temporarily unable to urinate, and bowel sounds are absent. In the space provided, identify the division of the autonomic nervous system that is affected by anesthesia.

_____

Developmental Aspects of the Nervous System

37. Complete the following statements by inserting your responses in the answer blanks.

_____ 1.

_____ 2.

_____ 3.

_____ 4.

_____ 5.

_____ 6.

_____ 7.

_____ 8.

Body temperature regulation is a problem in premature infants because the __(1)__ is not yet fully functional. Cerebral palsy involves crippling neuromuscular problems. It usually is a result of a lack of __(2)__ to the infant's brain during delivery. Normal maturation of the nervous system occurs in a __(3)__ direction, and fine control occurs much later than __(4)__ muscle control. The sympathetic nervous system becomes less efficient as aging occurs, resulting in an inability to prevent sudden changes in __(5)__ when abrupt changes in position are made. The usual cause of decreasing efficiency of the nervous system as a whole is __(6)__ . A change in intellect due to a gradual decrease in oxygen delivery to brain cells is called __(7)__ ; death of brain neurons, which results from a sudden cessation of oxygen delivery, is called a __(8)__ .

The Incredible Journey: A Visualization Exercise for the Nervous System

You climb on the first cranial nerve you see . . .

38. Where necessary, complete statements by inserting the missing words in the answer blanks.

_____ 1.

_____ 2.

_____ 3.

_____ 4.

Nervous tissue is quite densely packed, and it is difficult to envision strolling through its various regions. Imagine instead that each of the various functional regions of the brain has a computerized room in which you might observe what occurs in that particular area. Your assignment is to determine where you are at any given time during your journey through the nervous system.

5. _____

6. _____

7. _____

8. _____

9. _____

10. _____

11. _____

12. _____

You begin your journey after being miniaturized and injected into the warm pool of cerebrospinal fluid in your host's fourth ventricle. As you begin your stroll through the nervous tissue, you notice a huge area of branching white matter overhead. As you enter the first computer room you hear an announcement through the loudspeaker, "The pelvis is tipping too far posteriorly. Please correct. We are beginning to fall backward and will soon lose our balance." The computer responds immediately, decreasing impulses to the posterior hip muscles and increasing impulses to the anterior thigh muscles. "How is that, proprioceptor 1?" From this information, you determine that your first stop is __(1)__.

At the next computer room, you hear, "Blood pressure to head is falling; increase sympathetic nervous system stimulation of the blood vessels." Then, as it becomes apparent that your host has not only stood up but is going to run, you hear "Increase rate of impulses to the heart and respiratory muscles. We are going to need more oxygen and a faster blood flow to the skeletal muscles of the legs." You recognize that this second stop must be the __(2)__.

Computer room 3 presents a problem. There is no loudspeaker here. Instead, incoming messages keep flashing across the wall, giving only bits and pieces of information. "Four hours since last meal: stimulate appetite center. Slight decrease in body temperature: initiate skin vasoconstriction. Mouth dry: stimulate thirst center. Oh, a stroke on the arm: stimulate pleasure center." Looking at what has been recorded here—appetite, temperature, thirst, and pleasure—you conclude that this has to be the __(3)__.

Continuing your journey upward toward the higher brain centers, finally you are certain that you have reached the cerebral cortex. The first center you visit is quiet, like a library with millions of "encyclopedias" of facts and recordings of past input. You conclude that this must be the area where __(4)__ are stored, and that you are in the __(5)__ lobe. The next stop is close by. As you enter the computer center, you once again hear a loudspeaker: "Let's have the motor instructions to say tintinnabulation. Hurry, we don't want them to think we're tongue-tied." This area is obviously __(6)__. Your final stop in the cerebral cortex is a very hectic center. Electrical impulses are traveling back and forth between giant neurons, sometimes in different directions and sometimes back and forth between a small number of neurons. Watching intently, you try to make some sense out of these interactions, and suddenly realize that this *is* what is happening here. The neurons *are* trying to make some sense out of something, and this helps you decide that this must be the brain area where __(7)__ occurs in the __(8)__ lobe.

You hurry out of this center and retrace your steps back to the cerebrospinal fluid, deciding en route to observe a cranial nerve. You decide to pick one randomly and follow it to the organ it serves. You climb on to the first cranial nerve you see and slide down past the throat. Picking up speed, you quickly pass the heart and lungs and see the stomach and small intestine coming up fast. A moment later you land on the stomach and now you know that this wandering nerve has to be the __(9)__. As you look upward, you see that the nerve is traveling almost straight up and that you'll have to find an alternative route back to the cerebrospinal fluid. You begin to walk posteriorly until you find a spinal nerve, which you follow until you reach the vertebral column. You squeeze between two adjacent vertebrae to follow the nerve to the spinal cord. With your pocket knife you cut away the tough connective tissue covering the cord. Thinking that the __(10)__ covering deserves its name, you finally manage to cut an opening large enough to get through and you return to the warm bath of cerebrospinal fluid that it encloses. At this point you are in the   **(11)** , and from here you swim upward until you get to the lower brain stem. Once there, it should be an easy task to find the holes leading into the   **(12)**   ventricle, where your journey began.

Special Senses

The body's sensory receptors react to stimuli or changes occurring within the body and in the external environment. When triggered, these receptors send nerve impulses along afferent pathways to the brain for interpretation. Thus, these specialized parts of the nervous system allow the body to assess and adjust to changing conditions so that homeostasis may be maintained.

The minute receptors of general sensation that react to touch—pressure, pain, temperature changes, and muscle tension—are widely distributed in the body. In contrast, receptors of the special senses—sight, hearing, equilibrium, smell, and taste—tend to be localized and in many cases are quite complex.

The structure and function of the special sense organs are the subjects of the student activities in this chapter.

Eye and Vision

1. Complete the following statements by inserting your responses in the answer blanks.

   _____ 1.

   _____ 2.

   _____ 3.

   _____ 4.

   Attached to the eyes are the **(1)** muscles that allow us to direct our eyes toward a moving object. The anterior aspect of each eye is protected by the **(2)** , which have eyelashes projecting from their edges. Associated with the lashes are oil-secreting glands called **(3)** that help to lubricate the eyes. An inflammation of one of these glands is called a **(4)** .

2. Trace the pathway that the secretion of the lacrimal glands takes from the surface of the eye by assigning a number to each structure. (Note that #1 will be *closest* to the lacrimal gland.)

   _____ **1.** Lacrimal sac   _____ **3.** Nasolacrimal duct

   _____ **2.** Nasal cavity   _____ **4.** Lacrimal ducts

3. Identify each of the eye muscles indicated by leader lines in Figure 8-1. Color each muscle a different color. Then, in the blanks below, indicate the eye movement caused by each muscle.

   1. Superior rectus _____

   2. Inferior rectus _____

   3. Superior oblique _____

   4. Lateral rectus _____

   5. Medial rectus _____

   6. Inferior oblique _____

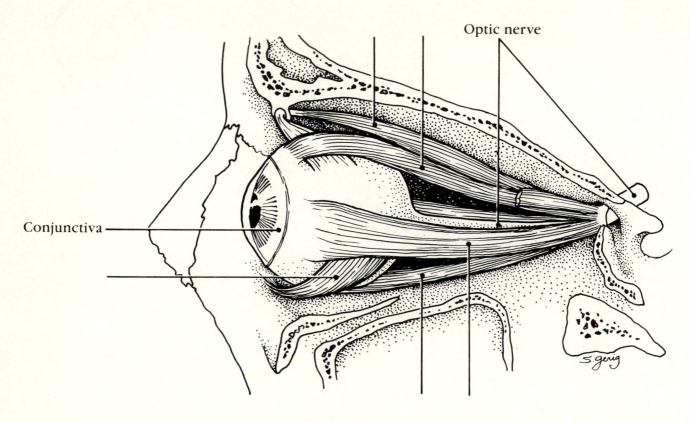

Optic nerve

Conjunctiva

Figure 8-1

4. Three accessory eye structures contribute to the formation of tears and/or aid in lubricating the eyeball. In the table, name each structure and then name its major secretory product. Indicate which of the secretions has antibacterial properties by circling that response.

Accessory eye structures	Secretory product
1.	
2.	
3.	

5. Match the terms provided in Column B with the appropriate descriptions in Column A. Insert the correct letter response or corresponding term in the answer blanks.

Column A	Column B
_____ **1.** Light bending	**A.** Accommodation
_____ **2.** Ability to focus for close vision (under 20 feet)	**B.** Accommodation pupillary reflex
_____ **3.** Normal vision	
_____ **4.** Inability to focus well on close objects; farsightedness	**C.** Astigmatism
	D. Cataract
_____ **5.** Reflex-constriction of pupils when they are exposed to bright light	**E.** Convergence
	F. Emmetropia
_____ **6.** Clouding of the lens, resulting in loss of sight	
_____ **7.** Nearsightedness	**G.** Glaucoma
	H. Hyperopia
_____ **8.** Blurred vision, resulting from unequal curvatures of the lens or cornea	**I.** Myopia
_____ **9.** Condition of increasing pressure inside the eye, resulting from blocked drainage of aqueous humor	**J.** Night blindness
	K. Photopupillary reflex
_____ **10.** Medial movement of the eyes during focusing on close objects	**L.** Refraction
_____ **11.** Reflex constriction of the pupils when viewing close objects	
_____ **12.** Inability to see well in the dark; often a result of vitamin A deficiency	

6. The intrinsic eye muscles are under the control of which division of the nervous system? Circle the correct response.

   1. Autonomic nervous system   **2.** Somatic nervous system

7. Complete the following statements by inserting your responses in the answer blanks.

_____ **1.**

_____ **2.**

_____ **3.**

_____ **4.**

_____ **5.**   _____ **6.**

A **(1)** lens, like that of the eye, produces an image that is upside down and reversed from left to right. Such an image is called a **(2)** image. In farsightedness, the light is focused **(3)** the retina. The lens used to treat farsightedness is a **(4)** lens. In nearsightedness, the light is focused **(5)** the retina; it is corrected with a **(6)** lens.

8. Using key choices, identify the parts of the eye described in the following statements. Insert the correct term or letter response in the answer blanks.

KEY CHOICES:

A.	Aqueous humor	**F.** ◯	Fovea centralis	**K.** ◯	Sclera
B.	Canal of Schlemm	**G.** ◯	Iris	**L.**	Suspensory ligaments
C. ◯	Choroid coat	**H.** ◯	Lens	**M.**	Vitreous humor
D. ◯	Ciliary body	**I.** ◯	Optic disk		
E. ◯	Cornea	**J.** ◯	Retina		

_____ **1.** Attaches the lens to the ciliary body

_____ **2.** Fluid that fills the anterior chamber of the eye; provides nutrients to the lens and cornea

_____ **3.** The "white" of the eye

_____ **4.** Area of retina that lacks photoreceptors; the blind spot

_____ **5.** Contains muscle that controls the shape of the lens

_____ **6.** Nutritive (vascular) tunic of the eye

_____ **7.** Drains the aqueous humor of the eye

_____ **8.** Tunic, containing the rods and cones

_____ **9.** Gel-like substance filling the posterior cavity of the eyeball; helps to reinforce the eyeball

_____ **10.** Heavily pigmented tunic that prevents light scattering within the eye

_____ **11.** _____ **12.** Smooth muscle structures (intrinsic eye muscles)

_____ **13.** Area of acute or discriminatory vision

_____ **14.** _____ **15.** _____ **16.** _____ **17.** Refractory media of the eye

_____ **18.** Anteriormost part of the sclera—your "window on the world"

_____ **19.** Tunic composed of tough, white fibrous connective tissue

9. Using the key choice terms given in Exercise 8, identify the structures indicated by leader lines on the diagram of the eye in Figure 8-2. Select different colors for all structures provided with a color-coding circle in Exercise 8, and then use them to color the coding circles and corresponding structures in the figure.

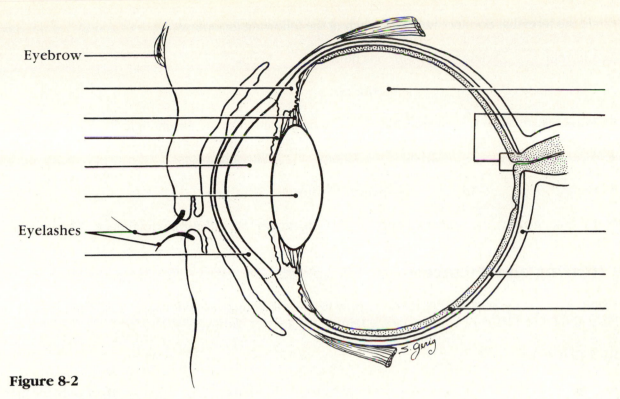

Eyebrow ———

Eyelashes ———

Figure 8-2

10. In the following table circle the correct word under the vertical headings that describes events occurring within the eye during close and distant vision.

Vision	Ciliary muscle		Lens convexity		Degree of light refraction	
1. Distant	Relaxed	Contracted	Increased	Decreased	Increased	Decreased
2. Close	Relaxed	Contracted	Increased	Decreased	Increased	Decreased

11. Name in sequence the neural elements of the visual pathway, beginning with the retina and ending with the optic cortex.

Retina ⟶ _____ ⟶ _____ ⟶ _____

Synapse in thalamus ⟶ _____ ⟶ Optic cortex

12. Complete the following statements by inserting your responses in the answer blanks.

_____ 1.

_____ 2.

_____ 3.

_____ 4.

_____ 5.

_____ 6.

There are __(1)__ varieties of cones. One type responds most vigorously to __(2)__ light, another to __(3)__ light, and still another to __(4)__ light. The ability to see intermediate colors such as purple results from the fact that more than one cone type is being stimulated __(5)__ . Lack of all color receptors results in __(6)__ . Because this condition is sex linked, it occurs more commonly in __(7)__ . Black and white, or dim light, vision is a function of the __(8)__ .

7. _____ 8.

13. Circle the term that does not belong in each of the following groupings.

    1. Choroid   Sclera   Vitreous humor   Retina

    2. Ciliary body   Iris   Superior rectus

    3. Pupil constriction   Far vision   Acommodation   Bright light

    4. Mechanoreceptors   Rods   Cones   Photoreceptors

    5. Ciliary body   Iris   Suspensory ligaments   Lens

    6. Lacrimal gland   Lacrimal duct   Conjunctiva   Nasolacrimal duct

Ear: Hearing and Balance

14. Using key choices, select the terms that apply to the following descriptions.
    Place the correct letter or corresponding term in the answer blanks.

    KEY CHOICES:

    A. Anvil (incus)   E. External auditory canal   I. Pinna   M. Tympanic membr:

    B. Cochlea   F. Hammer (malleus)   J. Round window   N. Vestibule

    C. Endolymph   G. Oval window   K. Semicircular canals

    D. Eustachian tube   H. Perilymph   L. Stirrup (stapes)

    _____ **1.** ____ **2.** ____ **3.** Structures composing the outer ear.

    _____ **4.** ____ **5.** ____ **6.** Structures composing the bony or osseous labyrinth

    _____ **7.** ____ **8.** ____ **9.** Collectively called the ossicles

    _____ **10.** ____ **11.** Ear structures not involved with hearing

    _____ **12.** Allows pressure in the middle ear to be equalized with the atmospheric pressure

    _____ **13.** Vibrates as sound waves hit it; transmits the vibrations to the ossicles

    _____ **14.** Contains the organ of Corti

    _____ **15.** Connects the nasopharynx and the middle ear

    _____ **16.** ____ **17.** Contain receptors for the sense of equilibrium

    _____ **18.** Transmits the vibrations from the stirrup to the fluid in the inner ear

    _____ **19.** Fluid contained inside the membranous labyrinth

    _____ **20.** Fluid contained within the osseous labyrinth, which bathes the membranous labyrinth

15. Figure 8-3 is a diagram of the ear. Use anatomic terms (as needed) from key choices in Exercise 14 to correctly identify all structures in the figure provided with leader lines. Color all external ear structures yellow; color the ossicles red; color the equilibrium areas of the inner ear green; and color the inner ear structures involved with hearing blue.

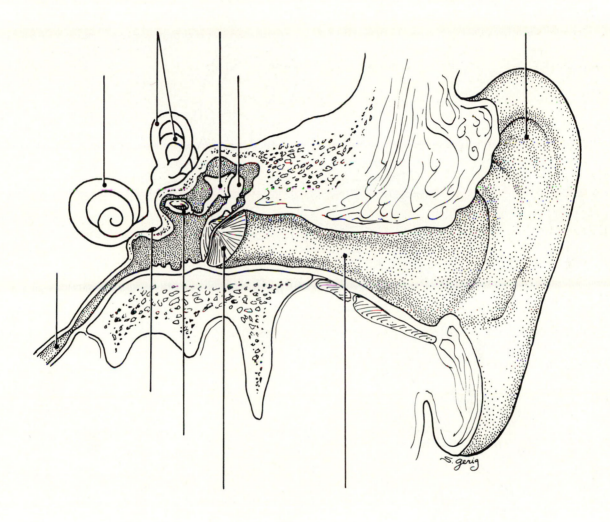

Figure 8-3

16. Sound waves hitting the eardrum set it into vibration. Trace the pathway through which vibrations and fluid currents travel to finally stimulate the hair cells in the organ of Corti. Name the appropriate ear structures in their correct sequence and insert your responses in the answer blanks.

Eardrum ⟶ _____ ⟶ _____ ⟶

_____ ⟶ Oval window ⟶ _____ ⟶

_____ ⟶ _____ ⟶ Hair cells

17. Figure 8-4 is a view of the structures of the membranous labyrinth. Correctly identify the following major areas of the labyrinth on the figure: *membranous semicircular canals, saccule* and *utricle,* and the *cochlear duct.* Next, correctly identify each of the receptor types shown in enlarged views (organ of Corti, crista ampullaris, and macula). Finally, using terms from the key choices below, identify all receptor structures provided with leader lines. (Some of these terms may need to be used more than once.)

Figure 8-4

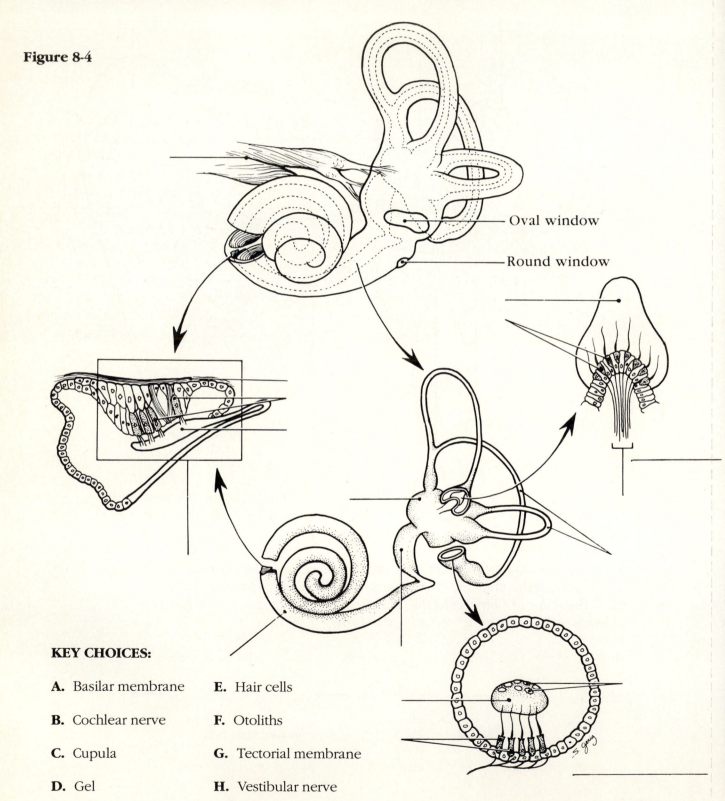

Oval window

Round window

KEY CHOICES:

A. Basilar membrane   **E.** Hair cells

B. Cochlear nerve   **F.** Otoliths

C. Cupula   **G.** Tectorial membrane

D. Gel   **H.** Vestibular nerve

18. Complete the following statements on the functioning of the static and dynamic equilibrium receptors by inserting the letter or term of the key choices in the answer blanks.

KEY CHOICES:

A. Angular/rotatory   **E.** Gravity   **I.** Semicircular canals

B. Cupula   **F.** Perilymph   **J.** Static

C. Dynamic   **G.** Proprioception   **K.** Utricle

D. Endolymph   **H.** Saccule   **L.** Vision

_____ 1.

_____ 2.

_____ 3.

_____ 4.

_____ 5.

_____ 6.

_____ 7.

_____ 8.

_____ 9.

_____ 10.

_____ 11.

The receptors for __(1)__ equilibrium are found in the crista ampullaris of the __(2)__. These receptors respond to changes in __(3)__ motion. When motion begins, the __(4)__ fluid lags behind and the __(5)__ is bent, which excites the hair cells. When the motion stops suddenly, the fluid flows in the opposite direction and again stimulates the hair cells. The receptors for __(6)__ equilibrium are found in the maculae of the __(7)__ and __(8)__. These receptors report the position of the head in space. Tiny stones found in a gel overlying the hair cells roll in response to the pull of __(9)__. As they roll, the gel moves and tugs on the hair cells, exciting them. Besides the equilibrium receptors of the inner ear, the senses of __(10)__ and __(11)__ are also important in helping to maintain equilibrium.

19. Indicate whether the following conditions relate to conduction deafness (*C*) or sensorineural (central) deafness (*S*). Place the correct letter choice in the answer blanks.

_____ **1.** Can result from the fusion of the ossicles

_____ **2.** Can result from damage to the cochlear nerve

_____ **3.** Sound is heard in one ear but not in the other, during both bone and air conduction

_____ **4.** Often improved by a hearing aid

_____ **5.** Can result from otitis media

_____ **6.** Can result from excessive earwax or a perforated eardrum

_____ **7.** Can result from a blood clot in the auditory cortex of the brain

20. List three things about which a person with equilibrium problems might complain. Place your responses in the answer blanks.

_____ , _____ , and _____

21. Circle the term that does not belong in each of the following groupings.

    1. Hammer   Anvil   Pinna   Stirrup

    2. Tectorial membrane   Crista ampullaris   Semicircular canals   Cupula

    3. Gravity   Angular motion   Sound waves   Rotation

    4. Utricle   Saccule   Eustachian tube   vestibule

    5. Vestibular nerve   Optic nerve   Cochlear nerve   Vestibulocochlear nerve

    6. Crista ampullaris   Maculae   Retina   Proprioceptors   Pressure receptors

Chemical Senses: Smell and Taste

22. Complete the following statements by inserting your responses in the answer blanks.

_____ **1.**

_____ **2.**

_____ **3.**

_____ **4.**

_____ **5.**

_____ **6.**

_____ **7.**

_____ **8.**

_____ **9.**

_____ **10.**

_____ **11.**

_____ **12.**

_____ **13.**

_____ **14.**

Three cranial nerves involved in transmitting impulses for the sense of taste are the __(1)__ , __(2)__ , and __(3)__ . Impulses for the sense of smell are transmitted by the __(4)__ nerve. The receptors for smell are poorly located in the __(5)__ of the nasal passages. The act of __(6)__ increases the sensation, because it brings more air into contact with the receptors. The receptors for taste are found in clusterlike areas called __(7)__ , most of which are located on the sides of __(8)__ or __(9)__ papillae. The four basic taste sensations are __(10)__ , __(11)__ , __(12)__ , and __(13)__ . The most protective receptors are thought to be those that respond to __(14)__ substances. When nasal passages are congested, the sense of taste is decreased. This indicates that much of what is considered taste actually depends on the sense of __(15)__ . It is impossible to taste substances with a __(16)__ tongue, because foods must be dissolved (or in solution) to excite the taste receptors. The sense of smell is closely tied to the emotional centers of the brain (limbic region), and many odors bring back __(17)__ .

_____ **15.**

_____ **16.**

_____ **17.**

23. On Figure 8-5(A), label the two types of tongue papillae containing taste buds. Then, using four different colors and appropriate labels, identify the areas of the tongue that are the predominant sites of sweet, salt, sour, and bitter receptors. On Figure 8-5(B) color the taste buds green. On Figure 8-5(C) color the gustatory cells red and the supporting cells blue. Add appropriate labels to the leader lines provided to identify the *taste pore* and *cilia* of the gustatory cells.

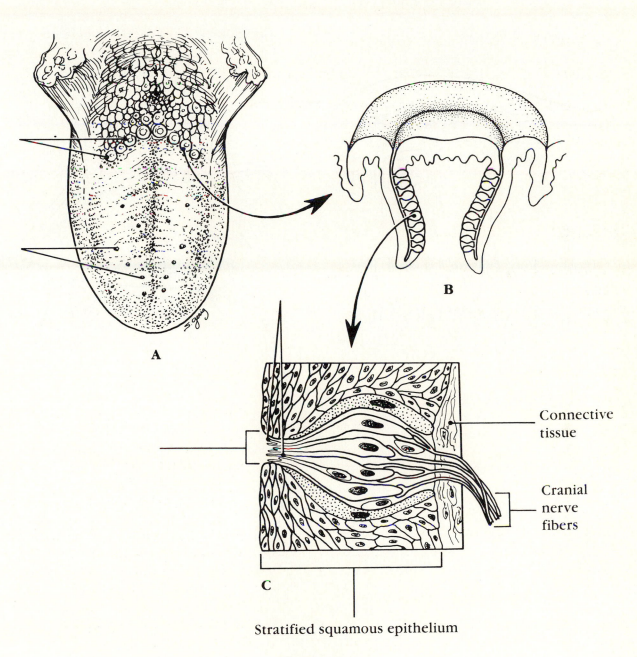

A

B

C

Connective tissue

Cranial nerve fibers

Stratified squamous epithelium

Figure 8-5

24. Figure 8-6 illustrates the site of the olfactory epithelium in the nasal cavity (part A is an enlarged view of the olfactory receptor area). Select different colors to identify the structures listed below and use them to color the coding circles and corresponding structures in the illustration. Then add a label and leader line to identify the olfactory "hairs" and add arrows to indicate the direction of impulse transmission.

○ Olfactory neurons (receptor cells)   ○ Olfactory bulb

○ Supporting cells   ○ Cribriform plate of the ethmoid bone

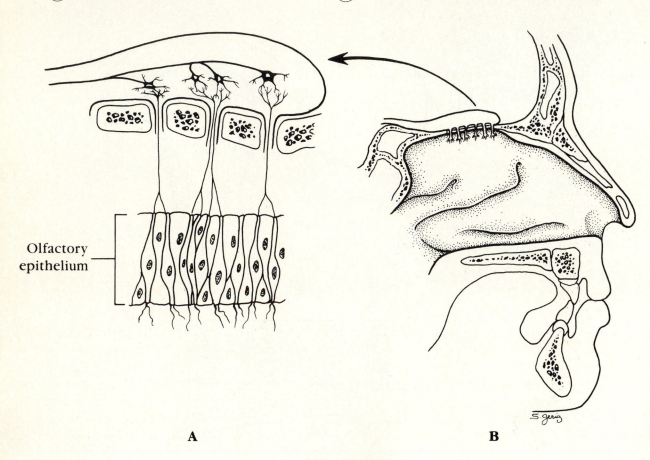

A   B

Figure 8-6

25. Circle the term that does not belong in each of the following groupings.

    1. Sweet   Musky   Sour   Bitter   Salty

    2. Bipolar neuron   Epithelial cell   Olfactory receptor   Ciliated

    3. Gustatory cell   Taste pore   Papillae   Neuron

    4. Vagus nerve   Facial nerve   Glossopharyngeal nerve   Olfactory nerve

    5. Olfactory receptor   High sensitivity   Variety of stimuli   Four receptor types

    6. Sugars   Sweet   Saccharine   Metal ions   Amino acids

    7. Taste buds   Olfactory receptors   Photoreceptors   Pain receptors

Developmental Aspects of the Special Senses

26. Complete the following statements by inserting your responses in the answer blanks.

    _____ 1.

    _____ 2.

    _____ 3.

    _____ 4.

    _____ 5.

    _____ 6.

    _____ 7.

    _____ 8.

    _____ 9.

    The special sense organs are actually part of the __(1)__ and are formed very early in the embryo. Maternal infections, particularly __(2)__, may cause both deafness and __(3)__ in the developing child. Of the special senses, the sense of __(4)__ requires the most learning or takes longest to mature. All infants are __(5)__, but generally by school age emmetropic vision has been established. Beginning sometime after the age of 40, the eye lenses start to become less __(6)__ and cannot bend properly to refract the light. As a result, a condition of farsightedness, called __(7)__, begins to occur. __(8)__, a condition in which the lens becomes hazy or discolored, is a frequent cause of blindness. In old age, a gradual hearing loss, called __(9)__, occurs. A declining efficiency of the chemical senses is also common in the elderly.

The Incredible Journey:
A Visualization Exercise
for the Special Senses

*You. . . see a discontinuous sea of glistening, white
rock slabs. . .*

27. Where necessary, complete statements by inserting the missing words in the
answer blanks.

_____ 1.

_____ 2.

_____ 3.

_____ 4.

_____ 5.

_____ 6.

_____ 7.

_____ 8.

_____ 9.

_____ 10.

_____ 11.

_____ 12.

_____ 13.

_____ 14.

_____ 15.

Your present journey will take you through your host's inner
ear to observe and document what you have learned about
how hearing and equilibrium receptors work.

This is a very tightly planned excursion. Your host has been
instructed to move his head at specific intervals and will be
exposed to various sounds so that you can make specific
observations. For this journey you are miniaturized and
injected into the bony cavity of the inner ear, the **(1)** , and
are to make your way through its various chambers in a
limited amount of time.

Your first observation is that you are in a warm sea of **(2)**
in the vestibule. To your right are two large sacs, the **(3)**
and **(4)** . You swim over to one of these membranous
sacs, cut a small semicircular opening in the wall, and
wiggle through. Since you are able to see very little in the
dim light, you set out to explore this area more fully. As you
try to move, however, you find that your feet are embedded
in a thick, gluelike substance. The best you can manage is
slow-motion movements through this **(5)** .

It is now time for your host's first scheduled head
movement. Suddenly your world tips sharply sideways. You
hear a roar (rather like an avalanche) and look up to see a
discontinuous sea of glistening, white rock slabs sliding
toward you. You protect yourself from these **(6)** by
ducking down between the hair cells that are bending
vigorously with the motion of the rocks. Now that you have
seen and can document the operation of a **(7)** , a sense
organ of **(8)** equilibrium, you quickly back out through the hole you made.

Keeping in mind the schedule and the fact that it is nearly time for your host to be exposed to
tuning forks, you swim quickly to the right, where you see what looks like the opening of a cave
with tall seaweed waving gently in the current. Abruptly, as you enter the cave, you find that you
are no longer in control of your movements, but instead are swept along in a smooth undulating
pattern through the winding passageway of the cave, which you now know is the cavity of the
(9) . As you move up and down with the waves you see hair cells of the **(10)** , the sense
organ for **(11)** , being vigorously disturbed below you. Flattening yourself against the chamber

wall to prevent being carried further by the waves, you wait for the stimulus to stop. Meanwhile you are delighted by the electrical activity of the hair cells below you. As they depolarize to send impulses along the __(12)__ nerve, the landscape appears to be alive with fireflies.

Now that you have witnessed the events for this particular sense receptor, you swim back through the vestibule toward your final observation area at the other end of the bony chambers. You recognize that your host is being stimulated again because of the change in fluid currents, but since you are not close to any of the sensory receptors, you are not sure just what the stimulus is. Then, just before you, three dark openings appear, the __(13)__ . You swim into the middle opening and see a strange structure that looks like the brush end of an artist's paintbrush; you swim upward and establish yourself on the soft brushy top portion. This must be the __(14)__ , the sensory receptor for __(15)__ equilibrium. As you rock back and forth in the gentle currents, a sudden wave of fluid hits you. Clinging to the hairs as the fluid thunders past you, you realize that there will soon be another such wave in the opposite direction. You decide that you have seen enough of the special senses and head back for the vestibule to leave your host once again.

Endocrine System

The endocrine system, vital to homeostasis, plays an important role in regulating the activity of body cells. Acting through blood-borne chemical messengers, called hormones, the endocrine system organs orchestrate cellular changes that lead to growth and development, reproductive capability, and the physiological homeostasis of many body systems.

This chapter promotes understanding of the location of the various endocrine organs in the body, the general function of the various hormones, and the consequences of their hypersecretion or hyposecretion.

The Endocrine System and Hormone Function—An Overview

1. Complete the following description of the manner in which hormone binding influences cellular machinery by writing the missing words in the answer blanks.

_____ 1.

_____ 2.

_____ 3.

_____ 4.

_____ 5.

_____ 6.

_____ 7.

In one mechanism, the hormone directly activates the cell's genes. This mechanism is typical of __(1)__ hormones, which are lipid-soluble and can diffuse through the plasma membrane. Once in the cell, these hormones attach to __(2)__ in the nucleus. The result of gene activation is the synthesis of specific __(3)__ .

The cyclic AMP mechanism is a good example of a second messenger system. In this mechanism, the hormone, acting as the __(4)__ messenger, binds to target cell membrane receptors coupled to the enzyme __(5)__ . As a result, the enzyme is activated and catalyzes the conversion of intracellular __(6)__ to cyclic AMP. Cyclic AMP then acts as the __(7)__ messenger to initiate a cascade of reactions in the target cell.

2. Complete the following statements by choosing a response from the key choices. Record the key letter or corresponding term in the answer blanks.

KEY CHOICES:

A. All body cells	**H.** Hypothalamus	**O.** Releasing hormones
B. Altering activity	**I.** More rapid	**P.** Slower/more prolonged
C. Anterior pituitary	**J.** Negative feedback	**Q.** Steroid or amino acid-based
D. Circulatory system	**K.** Nerve impulses	**R.** Stimulating new or unusual activities
E. Hormonal	**L.** Nervous system	**S.** Sugar or protein
F. Hormones	**M.** Neural	**T.** Target cell(s)
G. Humoral	**N.** Receptors	

_____ 1.

_____ 2.

_____ 3.

_____ 4.

_____ 5.

_____ 6.

_____ 7.

_____ 8.

_____ 9.

_____ 10.

_____ 11.

_____ 12.

_____ 13.

_____ 14.

_____ 15.

_____ 16.

_____ 17.

_____ 18.

The endocrine system is a major controlling system in the body. Its means of control, however, is much __(1)__ than that of the __(2)__ , the other major body system that acts to maintain homeostasis. Perhaps the reason for this is that the endocrine system uses chemical messengers, called __(3)__ , instead of __(4)__ . These chemical messengers enter the blood and are carried throughout the body by the activity of the __(5)__ . __(6)__ do not respond, however, to endocrine system stimulation. Only those that have the proper __(7)__ on their cell membranes are activated by the chemical messengers. These responsive cells are called the __(8)__ of the various endocrine glands. Hormones promote homeostasis by __(9)__ of body cells rather than by __(10)__ . All hormones are __(11)__ molecules. The various endocrine glands are prodded to release their hormones by nerve fibers (a __(12)__ stimulus), or by other hormones (a __(13)__ stimulus), or by the presence of increased or decreased levels of various other substances in the blood (a __(14)__ stimulus). The secretion of most hormones is regulated by a __(15)__ system, in which increasing levels of that particular hormone "turn off" its stimulus. The __(16)__ is called the master endocrine gland because it regulates so many other endocrine organs. However, it is, in turn, controlled by the __(17)__ secreted by the __(18)__ .

Endocrine Organs—Location and Function

3. Name the hormone that best fits each of the following descriptions. Insert your responses in the answer blanks.

_____ **1.** Basal metabolic hormone

_____ **2.** Programs T lymphocytes

_____ **3.** Most important hormone, regulating the amount of calcium circulation in the blood; released when blood calcium levels drop

_____ **4.** Helps to protect the body during long-term stressful situations such as extended illness and surgery.

_____ **5.** Short-term stress hormone; aids in the fight-or-flight response; increases blood pressure and heart rate, for example

_____ **6.** Necessary if glucose is to be taken up by body cells

_____ **7.** _____ **8.** Regulate the function of another endocrine organ; four tropic

_____ **9.** _____ **10.** hormones

_____ **11.** Acts antagonistically to insulin; produced by the same endocrine organ

_____ **12.** Hypothalamic hormone important in regulating water balance

_____ **13.** _____ **14.** Regulate the ovarian cycle

_____ **15.** _____ **16.** Directly regulate the menstrual or uterine cycle

_____ **17.** Adrenal cortex hormone involved in regulating salt levels of body fluids

_____ **18.** _____ **19.** Necessary for milk production and ejection

4. Figure 9-1 is a diagram of the various endocrine organs of the body. Next to each letter on the diagram, write the name of the endocrine-producing organ (or area). Then select different colors for each and color the corresponding organs in the illustration. To complete your identification of the hormone-producing organs, name the organs (not illustrated) described in items J and K.

J. Small glands that ride "horseback" on the thyroid

_____

K. Endocrine-producing organ present only in pregnant women

_____

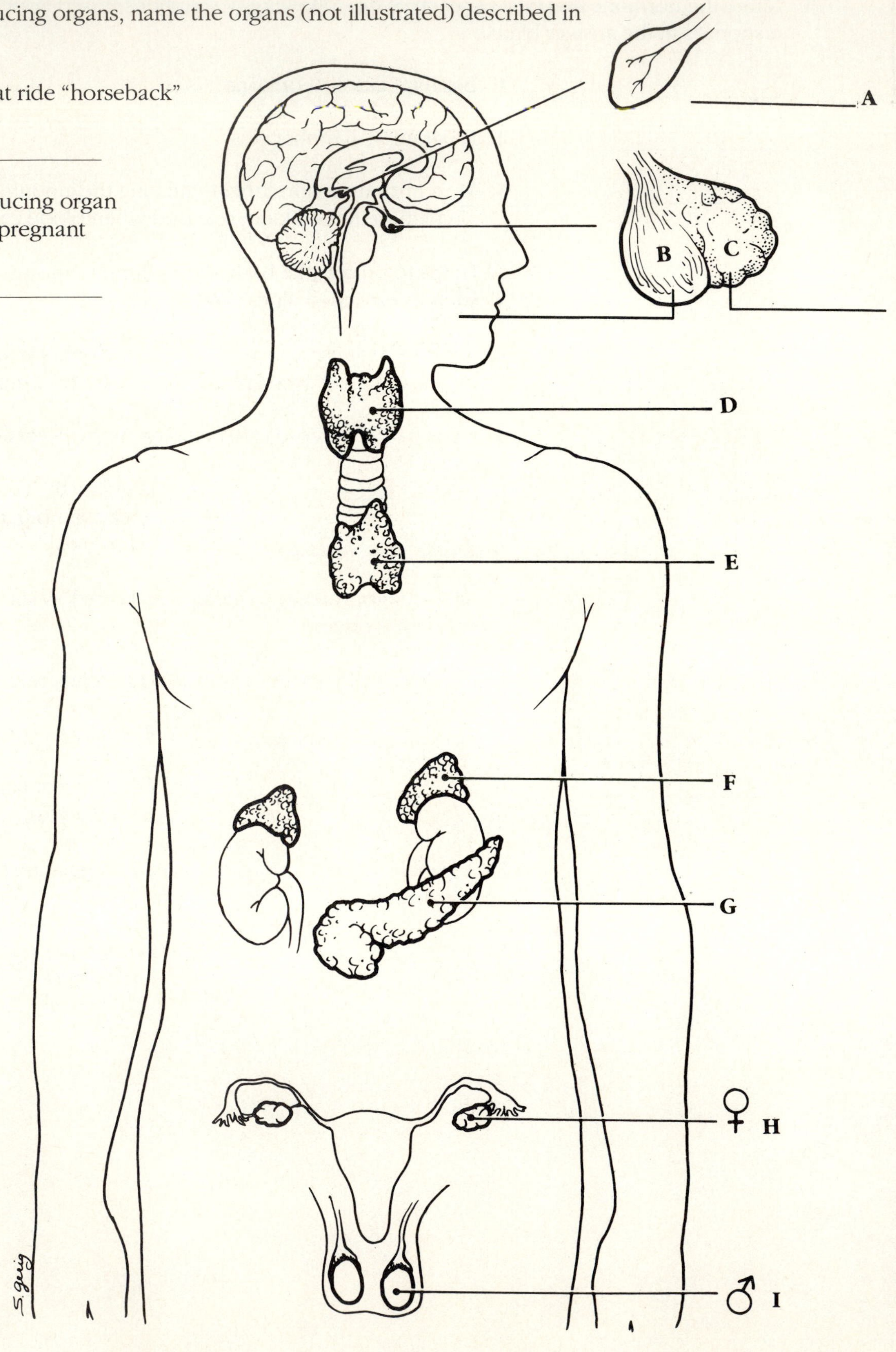

Figure 9-1

5. For each of the following hormones, indicate the organ (or organ part) producing or releasing the hormone by inserting the appropriate letters from Figure 9-1 in the answer blanks.

_____ **1.** ACTH

_____ **2.** ADH

_____ **3.** Aldosterone

_____ **4.** Cortisone

_____ **5.** Epinephrine

_____ **6.** Estrogen

_____ **7.** FSH

_____ **8.** Glucagon

_____ **9.** Insulin

_____ **10.** LH

_____ **11.** Melatonin

_____ **12.** Oxytocin

_____ **13.** Progesterone

_____ **14.** Prolactin

_____ **15.** PTH

_____ **16.** STH (growth hormone)

_____ **17.** Testosterone

_____ **18.** Thymosin

_____ **19.** Thyrocalcitonin

_____ **20.** Thyroxine

_____ **21.** TSH

6. Parathyroid hormone (PTH) has multiorgan effects. Indicate its effects on the organs listed below.

1. Kidneys _____

2. Intestine _____

3. Bones _____

Conditions Caused by Endocrine Organ Hypoactivity or Hyperactivity

7. Name the hormone that would be produced in *inadequate* amounts in the following conditions. Place your responses in the answer blanks.

_____ **1.** Sexual immaturity

_____ **2.** Tetany

_____ **3.** Excessive urination without high blood glucose levels; causes dehydration and tremendous thirst

_____ **4.** Goiter

_____ **5.** Cretinism; a type of dwarfism in which the individual retains childlike proportions and is mentally retarded

_____ **6.** Excessive thirst, high blood glucose levels, acidosis

_____ **7.** Abnormally small stature, normal proportions

_____ **8.** Miscarriage

_____ **9.** Lethargy, falling hair, low basal metabolic rate, obesity (myxedema in the adult)

8. Name the hormone that would be produced in *excessive* amounts in the following conditions. Place your responses in the answer blanks.

_____ 1. Lantern jaw; large hands and feet (acromegaly in the adult)

_____ 2. Bulging eyeballs, nervousness, increased pulse rate, weight loss (Graves' disease)

_____ 3. Demineralization of bones; spontaneous fractures

_____ 4. Cushing's syndrome—moon face, depression of the immune system

_____ 5. Abnormally large stature, relatively normal body proportions

_____ 6. Abnormal hairiness; masculinization

9. List the cardinal symptoms of diabetes mellitus, and provide the rationale for the occurrence of each symptom.

1. _____

_____

2. _____

_____

3. _____

_____

Developmental Aspects of the Endocrine System

10. Complete the following statements by inserting your responses in the answer blanks.

_____ 1.

_____ 2.

_____ 3.

_____ 4.

_____ 5.

_____ 6.

_____ 7.

Under ordinary conditions, the endocrine organs operate smoothly until old age. However, a __(1)__ in an endocrine organ may lead to __(2)__ of its hormones. A lack of __(3)__ in the diet may result in undersecretion of thyroxine. Later in life, a woman experiences a number of symptoms such as hot flashes and mood changes, which result from decreasing levels of __(4)__ in her system. This period of a woman's life is referred to as __(5)__ , and it results in a loss of her ability to __(6)__ . Because __(7)__ production tends to decrease in an aging person, adult-onset diabetes is common.

The Incredible Journey:
A Visualization Exercise
for the Endocrine System

*. . .you notice charged particles, shooting pell-mell
out of the bone matrix. . .*

11. Where necessary, complete statements by inserting the missing words in the
answer blanks.

_____ 1.

_____ 2.

_____ 3.

_____ 4.

_____ 5.

_____ 6.

_____ 7.

_____ 8.

_____ 9.

For this journey, you will be miniaturized and injected into a
vein of your host. Throughout the journey, you will be
traveling in the bloodstream. Your instructions are to record
changes in blood composition as you float along and to
form some conclusions as to why they are occurring (that is,
which hormone is being released).

Bobbing gently along in the slowly moving blood, you
realize that there is a sugary taste to your environment;
however, the sweetness begins to decrease quite rapidly. As
the glucose levels of the blood have just decreased,
obviously __(1)__ has been released by the __(2)__ , so that the
cells can take up glucose.

A short while later you notice that the depth of the blood in
the vein you are traveling in has diminished substantially. To
remedy this potentially serious situation, the __(3)__ will have
to release more __(4)__ , so the kidney tubules will reabsorb
more water. Within a few minutes the blood becomes much deeper; you wonder if the body is
psychic as well as wise.

As you circulate past the bones, you notice charged particles, shooting pell-mell out of the bone
matrix and jumping into the blood. You conclude that the __(5)__ glands have just released PTH,
because the __(6)__ levels have increased in the blood. As you continue to move in the
bloodstream, the blood suddenly becomes sticky sweet, indicating that your host must be nervous
about something. Obviously, his __(7)__ has released __(8)__ to cause this sudden increase in blood
glucose.

Sometime later, you become conscious of a humming activity around you, and you sense that the
cells are very busy. Obviously your host's __(9)__ levels are sufficient, since his cells are certainly
not sluggish in their metabolic activities. You record this observation and prepare to end this
journey.

Blood

Blood, the vital "life fluid" that courses through the body's blood vessels, provides the means by which the body's cells receive vital nutrients and oxygen and dispose of their metabolic wastes. As blood flows past the tissue cells, exchanges continually occur between the blood and the cells, so that vital activities are maintained.

In this chapter, the student has an opportunity to review the general characteristics of whole blood and plasma, to identify the various formed elements (blood cells), and to recall their functions. Blood groups, transfusion reactions, clotting, and various types of blood abnormalities are also considered.

Composition and Function

1. Complete the following description of the components of blood by writing the missing words in the answer blanks.

_____ 1.

_____ 2.

_____ 3.

_____ 4.

_____ 5.

_____ 6.

_____ 7.

_____ 8.

_____ 9.

_____ 10.

In terms of its tissue classification, blood is classified as a __(1)__ because it has living blood cells, called __(2)__, suspended in a nonliving fluid matrix called __(3)__. The "fibers" of blood only become visible during __(4)__.

If a blood sample is centrifuged, the heavier blood cells become packed at the bottom of the tube. Most of this compacted cell mass is composed of __(5)__, and the volume of blood accounted for by these cells is referred to as the __(6)__. The less dense __(7)__ rises to the top and constitutes about 45% of the blood volume. The so-called "buffy coat" composed of __(8)__ and __(9)__ is found at the junction between the other two blood elements. The buffy coat accounts for less than __(10)__ % of blood volume.

Blood is scarlet red in color when it is loaded with __(11)__; otherwise, it tends to be dark red.

_____ 11.

2. Using key choices, identify the cell type(s) or blood elements that fit the following descriptions. Insert the correct term or letter response in the spaces provided.

KEY CHOICES:

A. Red blood cell   **D.** Basophil   **G.** Lymphocyte

B. Megakaryocyte   **E.** Monocyte   **H.** Formed elements

C. Eosinophil   **F.** Neutrophil   **I.** Plasma

_____ **1.** Most numerous leukocyte

_____ **2.** _____ **3.** _____ **4.** Granular leukocytes

_____ **5.** Also called an erythrocyte; anucleate

_____ **6.** _____ **7.** Actively phagocytic leukocytes

_____ **8.** _____ **9.** Agranular leukocytes

_____ **10.** Fragments to form platelets

_____ **11.** (A) through (G) are examples of these

_____ **12.** Increases during allergy attacks

_____ **13.** Releases histamine during inflammatory reactions

_____ **14.** After originating in bone marrow, may be formed in lymphoid tissue

_____ **15.** Contains hemoglobin; therefore involved in oxygen transport

_____ **16.** Primarily water, noncellular; the fluid matrix of blood

_____ **17.** Increases in number during prolonged infections

_____ **18.** Least numerous leukocyte

_____ **19.** _____ **20.** _____ **21.** _____ **22.** _____ **23.** Also called white blood cells

3. Figure 10-1 depicts in incomplete form the erythropoietin mechanism for regulating the rate of erythropoiesis. Complete the statements with answer blanks, label structures with leader lines, and then choose colors (other than yellow) for the color coding circles and corresponding structures on the diagram. Color all arrows on the diagram yellow. Finally, indicate the normal life span of erythrocytes to complete this exercise.

    ◯   The kidney   ◯   Red bone marrow   ◯   RBCs

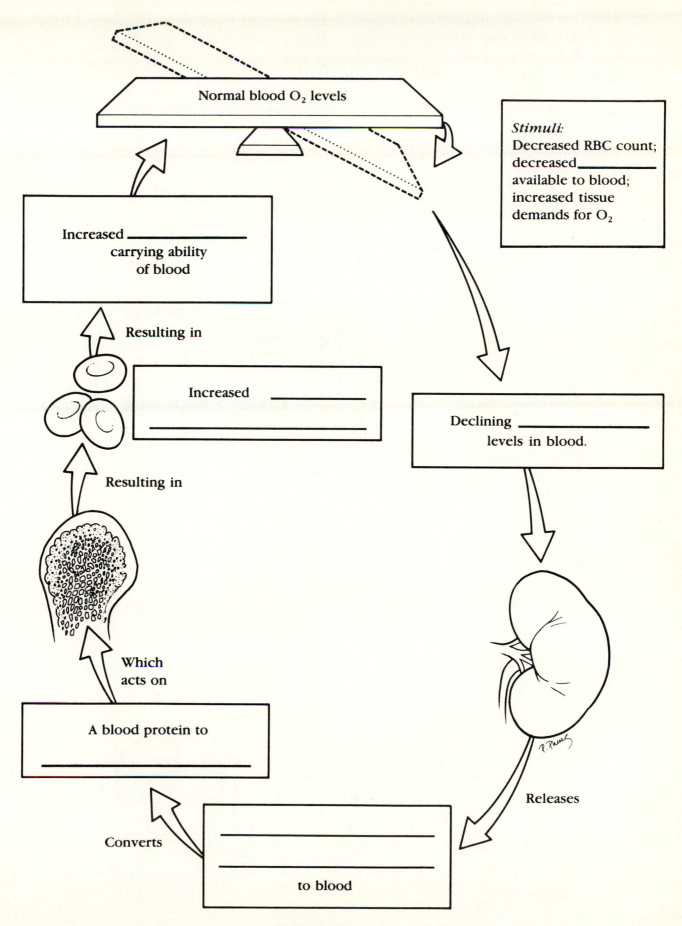

Normal blood O₂ levels

Stimuli:
Decreased RBC count;
decreased _____
available to blood;
increased tissue
demands for O₂

Increased _____
carrying ability
of blood

Resulting in

Increased _____
_____

Declining _____
levels in blood.

Resulting in

Which
acts on

A blood protein to
_____

Converts

_____
_____
to blood

Releases

Figure 10-1

4. Four leukocytes are diagrammed in Figures 10-2 to 10-5. First, follow directions (given below) for coloring each leukocyte as it appears when stained with Wright's stain. Then, identify each leukocyte type by writing in the correct name in the blank below the illustration.

Figure 10-2. Color the granules pale violet, the cytoplasm pink, and the nucleus dark purple.

Figure 10-3. Color the nucleus deep blue and the cytoplasm pale blue

Figure 10-4. Color the granules bright red, the cytoplasm pale pink, and the nucleus red/purple.

Figure 10-5. For this smallest white blood cell, color the nucleus deep purple/blue and the sparse cytoplasm pale blue.

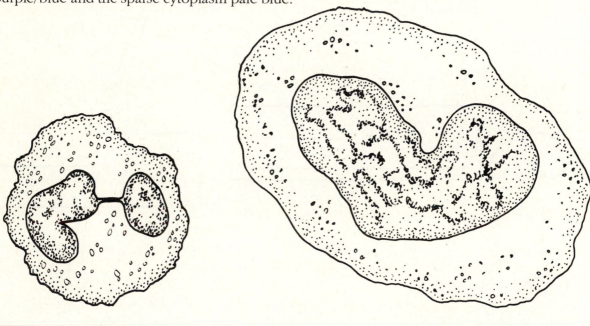

Figure 10-2   **Figure 10-3**

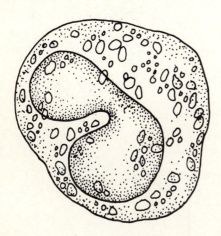

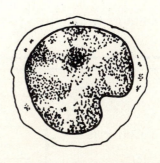

Figure 10-4   **Figure 10-5**

5. If the following statements are true, insert *T* in the answer blank. If any of the statements are false, correct the underlined word(s) by inserting the correct word in the answer blank.

_____   1. White blood cells (WBCs) move into and out of blood vessels by the process of chemotaxis.

_____   2. A count of white blood cells that provides information on the relative number of each WBC type is the total WBC count.

_____   3. When blood becomes too acid or too basic, both the respiratory system and the liver may be called into action to restore it to its normal pH range.

_____   4. The normal pH range of blood is 7.00 to 7.45.

_____   5. The cardiovascular system of an average adult contains approximately 4 liters of blood.

_____   6. Blood is circulated through the blood vessels by the pumping action of the heart.

_____   7. The universal ABO group donor is AB.

_____   8. The normal RBC count is 3.5–4.5 million/mm^3.

_____   9. Normal hemoglobin values are in the area of 42%–47% of the volume of whole blood.

_____   10. An anemia resulting from a decreased RBC number causes the blood to become more viscous.

_____   11. The most important natural body anticoagulant is histamine.

_____   12. Phagocytic agranular WBCs are eosinophils.

_____   13. The leukocytes particularly important in the immune response are monocytes.

_____   14. Normally, blood clots within 5–10 minutes.

_____   15. An abnormal increase in the number of red blood cells is leukocytosis.

_____   16. An abnormal decrease in the number of white blood cells is leukopenia.

6. Using key choices, correctly complete the following description of the blood-clotting process. Insert the key term or letter in the answer blanks.

KEY CHOICES:

A. Break   D. Fibrinogen   G. Serotonin

B. Erythrocytes   E. Platelets   H. Thrombin

C. Fibrin   F. Prothrombin   I. Thromboplastin

_____ 1.

_____ 2.

_____ 3.

_____ 4.

_____ 5.

_____ 6.

_____ 7.

_____ 8.

_____ 9.

Clotting begins when a __(1)__ occurs in a blood vessel wall. Almost immediately, __(2)__ cling to the blood vessel wall and release __(3)__ , which helps to decrease blood loss by constricting the vessel. __(4)__ is also released. This chemical substance causes __(5)__ to be converted to __(6)__ . Once present, thrombin acts as an enzyme to attach __(7)__ molecules together to form long, threadlike strands of __(8)__ , which then traps __(9)__ flowing by in the blood.

7. Circle the term that does not belong in each of the following groupings.

1. Erythrocytes   Lymphocytes   Monocytes   Eosinophils

2. Neutrophils   Monocytes   Basophils   Eosinophils

3. Histamine   Heparin   Basophil   Antibodies

4. Hemoglobin   Lymphocyte   Oxygen transport   Erythrocytes

5. Thrombocytes   Monocytes   Phagocytosis   Neutrophils

6. Thrombus   Aneurysm   Embolus   Clot

7. Increased hemoglobin   Decreased hemoglobin   Red blood cell lysis   Anemia

8. Plasma   Nutrients   Hemoglobin   Wastes

9. Myeloid stem cell   Lymphocyte   Monocyte   Basophil

Blood Groups and Transfusions

8. Correctly complete the following table concerning ABO blood groups.

Blood Type	Agglutinogens or antigens	Agglutinins or antibodies in plasma	Can donate blood to type	Can receive blood from type
1. Type A	A			
2. Type B		anti-A		
3. Type AB			AB	
4. Type O	none			

9. Ms. Pratt is claiming that Mr. X is the father of her child. Ms. Pratt's blood type is O negative. Her baby boy has type A positive blood. Mr. X's blood is typed and found to be B positive. Could he be the father of her child? _____ If not, what blood type would the father be expected to have? _____

10. When a person is given a transfusion of mismatched blood, a transfusion reaction occurs. Define the term "transfusion reaction" in the blanks provided here.

_____

_____

11. Correctly respond to questions referring to the following situation by placing your answers in the answer blanks.

    Mrs. Carlyle is pregnant for the first time. Her blood type is Rh negative, her husband is Rh positive, and their first child has been determined to be Rh positive. Ordinarily, the first such pregnancy causes no major problems, but baby Carlyle is born blue and cyanotic. What is this condition, a result of Rh

    incompatibility, called? _____

    Why is the baby cyanotic? _____

    _____

Since this is Mrs. Carlyle's first pregnancy, how can you account for the baby's

problem? (Think) _____

_____

Assume that baby Carlyle was born pink and healthy. What measures should be
taken to prevent the previous situation from happening in the second

pregnancy with an Rh positive baby? _____

_____

Developmental Aspects of Blood

12. Complete the following statements by inserting your responses in the answer
    blanks.

_____ **1.**   A fetus has a special type of hemoglobin, hemoglobin
                            __(1)__ , that has a particularly high affinity for oxygen. After
_____ **2.**   birth, the infant's fetal RBCs are rapidly destroyed and
                            replaced by hemoglobin A–containing RBCs. When the
_____ **3.**   immature infant liver cannot keep pace with the demands to
                            rid the body of hemoglobin breakdown products, the
_____ **4.**   infant's tissues become yellowed, or __(2)__ .

_____ **5.**   Genetic factors lead to several congenital diseases
                            concerning the blood. An anemia in which RBCs become
_____ **6.**   sharp and "log-jam" in the blood vessels under conditions of
                            low-oxygen tension in the blood is __(3)__ anemia. Bleeder's
_____ **7.**   disease, or __(4)__ , is a result of a deficiency of certain
                            clotting factors.
_____ **8.**
                            Diet is important to normal blood formation. Women are
_____ **9.**   particularly prone to __(5)__ -deficiency anemia because of
                            their monthly menses. A decreased efficiency of the gastric
                            mucosa makes elderly individuals particularly susceptible to
                            __(6)__ anemia as a result of a lack of intrinsic factor, which is
                            necessary for vitamin __(7)__ absorption. An important
                            problem in aged individuals is their tendency to form
                            undesirable clots, or __(8)__ . Both the young and the elderly
                            are at risk for cancer of the blood, or __(9)__ .

The Incredible Journey:
A Visualization Exercise for the Blood

*Once inside, you quickly make a slash in the vessel
lining . . .*

13. Where necessary, complete statements by inserting the missing words in the
answer blanks.

_____	1.
_____	2.
_____	3.
_____	4.
_____	5.
_____	6.
_____	7.
_____	8.
_____	9.
_____	10.
_____	11.
_____	12.
_____	13.
_____	14.
_____	15.
_____	16.
_____	17.
_____	18.
_____	19.

For this journey, you will be miniaturized and injected into
the external iliac artery and will be guided by a fluorescent
monitor into the bone marrow of the iliac bone. You will
observe and report events of blood cell formation, also
called **(1)** , seen there and then move out of the bone into
the circulation to initiate and observe the process of blood
clotting, also called **(2)** . Once in the bone marrow, you
observe as several large dark-nucleated stem cells, or **(3)** ,
begin to divide and produce daughter cells. To your right,
the daughter cells eventually formed have tiny cytoplasmic
granules and very peculiarly shaped nuclei that look like
small masses of nuclear material connected by thin strands
of nucleoplasm. You have just witnessed the formation of a
type of white blood cell called the **(4)** . You describe its
appearance and make a mental note to try to observe its
activity later. Meanwhile you can tentatively report that this
cell type functions as a **(5)** to protect the body.

At another site, daughter cells arising from the division of a
stem cell are difficult to identify initially. As you continue to
observe the cells, you see that they, in turn, divide.
Eventually some of their daughter cells eject their nuclei and
flatten out to assume a disk shape. You assume that the
kidneys must have released **(6)** , because those cells are
 (7) . That dark material filling their interior must be
 (8) , because those cells function to transport **(9)** in the
blood.

Now you turn your attention to the daughter cells being
formed by the division of another stem cell. They are small
round cells with relatively large round nuclei. In fact, their
cytoplasm is very sparse. You record your observation of the
formation of **(10)** . They do not remain in the marrow
very long after formation, but seem to enter the circulation

_____ 20.

_____ 21.

_____ 22.

_____ 23.

_____ 24.

_____ 25.

almost as soon as they are produced. Some of those cells produce **(11)** or act in other ways in the immune response. At this point, although you have yet to see the formation of **(12)** , **(13)** , **(14)** , or **(15)** , you decide to proceed into the circulation to make the blood-clotting observations.

You maneuver yourself into a small venule to enter the general circulation. Once inside, you quickly make a slash in the vessel lining, or **(16)** . Almost immediately what appear to be hundreds of jagged cell fragments swoop into the area and plaster themselves over the freshly made incision. You record that **(17)** have just adhered to the damaged site. As you are writing, your chemical monitor flashes the message, "vasoconstrictor substance released." You record that **(18)** has been released based on your observation that the vessel wall seems to be closing in. Peering out at the damaged site, you see that long ropelike strands are being formed at a rapid rate and are clinging to the site. You report that the **(19)** mesh is forming and is beginning to trap RBCs to form the basis of the **(20)** . Even though you do not have the equipment to monitor the intermediate steps of this process, you know that the platelets must have also released **(21)** , which then converted **(22)** to **(23)** . This second enzyme then joined the soluble **(24)** molecules together to form the network of strands you can see.

You carefully back away from the newly formed clot. You do *not* want to disturb the area because you realize that if the clot detaches, it might become a life-threatening **(25)** . Your mission here is completed, and you return to the entrance site.

Circulatory System

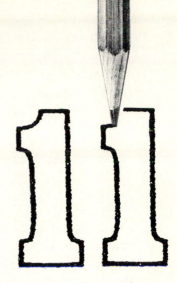

The major structures of the circulatory system, the heart and blood vessels, play a vital role in human physiology. The major function of the circulatory system is transportation. Using blood as the transport vehicle, the system carries nutrients, gases, wastes, antibodies, electrolytes, and many other substances to and from body cells. Its propulsive force is the contracting heart.

The anatomy and location of the heart and blood vessels and the important understandings of cardiovascular physiology (for example, cardiac cycle, ECG, and regulation of blood pressure) are the major topics of the student activities of this chapter. The structures and functioning of the lymphatic system, a pumpless adjunct to the cardiovascular structures, are also briefly considered.

Cardiovascular System: The Heart

1. The heart is called a double pump because it serves two circulations. Trace the flow of blood through both the pulmonary and systemic circulations by writing the missing terms in the answer blanks.

    1. _____

    2. _____

    3. _____

    4. _____

    5. _____

    6. _____

    7. _____   7. _____   9. _____   11.

    8. _____   8. _____   10. _____   12.

    From the right atrium through the tricuspid valve to the **(1)** , through the **(2)** valve to the pulmonary trunk to the right and left **(3)** , to the capillary beds of the **(4)** , to the **(5)** , to the **(6)** of the heart through the **(7)** valve, to the **(8)** through the **(9)** semilunar valve, to the systemic arteries, to the **(10)** of the body tissues, to the systemic veins, to the **(11)** and **(12)** , which enter the right atrium of the heart.

2. Figure 11-1 is an anterior view of the heart. Identify each numbered structure and write its name in the corresponding numbered space below the figure. Then, select different colors for each structure provided with a color-coding circle, and use them to color the coding circles and corresponding structures on the figure.

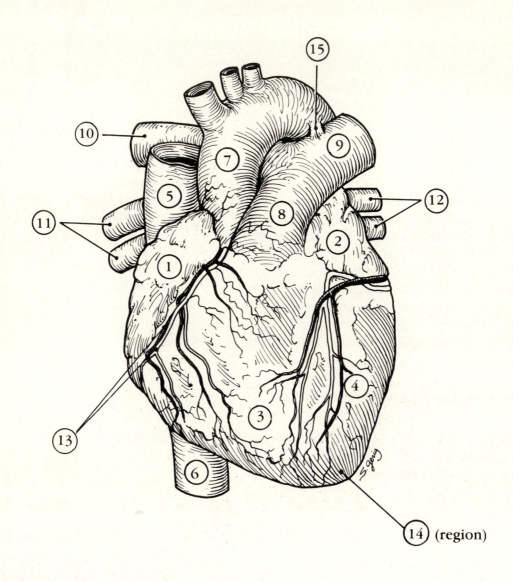

○ _____ 1.   ○ _____ 6.   ○ _____ 11.

○ _____ 2.   ○ _____ 7.   _____ 12.

○ _____ 3.   ○ _____ 8.   _____ 13.

○ _____ 4.   _____ 9.   _____ 14.

○ _____ 5.   _____ 10.   ○ _____ 15.

Figure 11-1

3. Complete the following statements by inserting your answers in the answer blanks.

_____ab_____ 1.

_____ 2.

_____ 3.

_____ 4.

_____ 5.

_____ 6.

_____ 7.

_____ 8.

_____ 9.

_____ 10.

_____ 11.

_____ 12.

_____ 13.

The heart is a cone-shaped muscular organ located within the __(1)__ . Its apex rests on the __(2)__ , and its base is at the level of the __(3)__ rib. The superior and inferior venae cavae empty venous blood into the __(4)__ . The pulmonary trunk, which divides into the right and left pulmonary arteries and carries blood to the lungs, arises from the __(5)__ . The four pulmonary veins empty into the __(6)__ . The coronary arteries that nourish the myocardium arise from the __(7)__ . The coronary sinus empties into the __(8)__ . The chamber that pumps blood into the aorta is the __(9)__ . Relative to the roles of the heart chambers, the __(10)__ are receiving chambers, whereas the __(11)__ are discharging chambers. The membrane that lines the heart and also forms the valve flaps is called the __(12)__ . The outermost layer of the heart is called the __(13)__ . The function of the fluid that fills the pericardial sac is to decrease __(14)__ during heart activity. The heart muscle, or myocardium, is composed of a specialized type of muscle tissue called __(15)__ .

_____ 14.

_____ 15.

4. The events of one complete heart beat are referred to as the cardiac cycle. Complete the following statements that describe these events. Insert your answers in the answer blanks.

_____ 1.

_____ 2.

_____ 3.

_____ 4.

_____ 5.

_____ 6.

_____ 7.

_____ 8.

_____ 9.

_____ 10.

The contraction of the ventricles is referred to as __(1)__ , and the period of ventricular relaxation is called __(2)__ . The monosyllables describing heart sounds during the cardiac cycle are __(3)__ . The first heart sound is a result of closure of the __(4)__ valves; closure of the __(5)__ valves causes the second heart sound. The heart chambers that have just been filled when you hear the first heart sound are the __(6)__ , and the chambers that have just emptied are the __(7)__ . Immediately after the second heart sound, the __(8)__ are filling with blood, and the __(9)__ are empty. Abnormal heart sounds, or __(10)__ , usually indicate valve problems.

5. Figure 11-2 is a diagram of the frontal section of the heart. Follow the instructions below to complete this exercise.

First, draw arrows to indicate the direction of blood flow through the heart. Draw the pathway of the oxygen-rich blood with red arrows, and trace the pathway of oxygen-poor blood with blue arrows.

Second, identify each of the elements of the intrinsic conduction system (numbers 1-5 on the figure) by inserting the appropriate terms in the blanks left of the figure. Then, indicate with green arrows the pathway that impulses take through this system.

Third, correctly identify each of the heart valves (numbers 6-9 on the figure) by inserting the appropriate terms in the blanks left of the figure and draw in and identify by name the cordlike structures that anchor the flaps of the atrioventricular (AV) valves.

Fourth, use the numbers from the figure to identify the structures described below. Place the numbers in the lettered answer blanks.

_____ **A.**   _____ **B.** Prevent backflow into the ventricles when the heart is relaxed

_____ **C.**   _____ **D.** Prevent backflow into the atria when the ventricles are contracting

_____ **E.** AV valve with three flaps

_____ **F.** AV valve with two flaps

_____ **G.** The pacemaker of the intrinsic conduction system

_____ **H.** The point in the intrinsic conduction system where the impulse is temporarily delayed

_____ **1.**

_____ **2.**

_____ **3.**

_____ **4.**

_____ **5.**

_____ **6.**

_____ **7.**

_____ **8.**

_____ **9.**

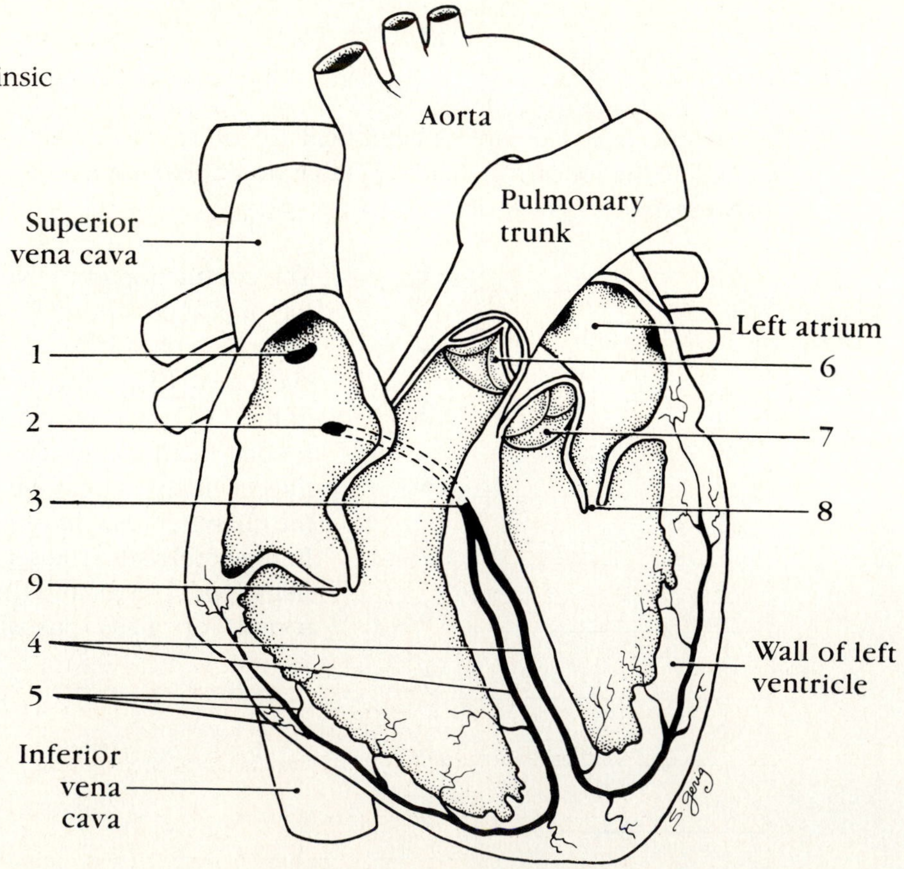

Figure 11-2

6. Match the terms provided in Column B with the statements given in Column A. Place the correct letter or term response in the answer blanks.

Column A   Column B

_____ 1. A recording of the elctrical activity of the heart   **A.** Angina pectoris

                                                                        B. Bradycardia

_____ 2. The period during which the atria are depolarizing   **C.** Electrocardiogram

_____ 3. The period during which the ventricles are repolarizing   **D.** Fibrillation

                                                                        E. Heart block

_____ 4. The period during which the ventricles are depolarizing, which precedes their contraction   **F.** P wave

                                                                        G. QRS wave

_____ 5. An abnormally slow heart beat, that is, below 60 beats per minute   **H.** T wave

_____ 6. A condition in which the heart is uncoordinated and useless as a pump   **I.** Tachycardia

_____ 7. An abnormally rapid heart beat, that is, over 100 beats per minute

_____ 8. Damage to the AV node, totally or partially releasing the ventricles from the control of the sinoatrial (SA) node

_____ 9. Chest pain, resulting from ischemia of the myocardium

7. A portion of an electrocardiogram is shown in Figure 11-3. On the figure identify the QRS complex, the P wave, and the T wave. Then, using a red pencil, bracket a portion of the recording equivalent to the length of one cardiac cycle. Using a blue pencil, bracket a portion of the recording in which the *ventricles* would be in diastole.

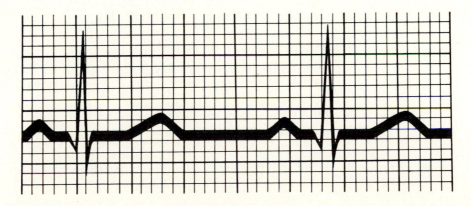

Figure 11-3

8. Complete the following statements relating to cardiac output by writing the missing terms in the answer blanks.

_____ 1.

_____ 2.

_____ 3.

_____ 4.

_____ 5.

_____ 6.

_____ 7.

_____ 8.

_____ 9.

_____ 10.

In the relationship CO = HR × SV, CO stands for __(1)__, HR stands for __(2)__, and SV stands for __(3)__. For the normal resting heart, the value of HR is __(4)__ and the value of SV is __(5)__. The normal average adult cardiac output, therefore, is __(6)__. The time for the entire blood supply to pass through the body is once each __(7)__.

According to Starling's Law of the heart, the critical factor that determines force of heartbeat, or __(8)__, is the degree of __(9)__ of the cardiac muscle just before it contracts. Consequently, the force of heartbeat can be increased by increasing the amount of __(10)__ returned to the heart.

9. Check (✔) all factors that lead to an increase in cardiac output by influencing either heart rate or stroke volume.

_____ **1.** Epinephrine   _____ **6.** Activation of the sympathetic nervous system

_____ **2.** Thyroxine   _____ **7.** Activation of the vagus nerves

_____ **3.** Hemorrhage   _____ **8.** Low blood pressure

_____ **4.** Fear   _____ **9.** High blood pressure

_____ **5.** Exercise   _____ **10.** Fever

10. For each of the following statements that is true, write *T* in the answer blank. For any false statements, correct the underlined term by writing the correct term in the answer blank.

_____ **1.** The resting heart rate is fastest in <u>adult</u> life.

_____ **2.** Because the heart of the highly trained athlete hypertrophies, its <u>stroke volume</u> decreases.

_____ **3.** If the <u>right</u> side of the heart fails, pulmonary congestion occurs.

_____ **4.** In <u>peripheral</u> congestion, the feet, ankles, and fingers become edematous.

_____ **5.** The pumping action of the healthy heart ordinarily maintains a balance between cardiac output and <u>venous return</u>.

11. Circle the term that does not belong in each of the following groupings.

    1. Pulmonary trunk   Vena cava   Right side of heart   Left side of heart

    2. QRS wave   T wave   P wave   Electrical activity of the ventricles

    3. AV valves closed   AV valves opened   Ventricular systole   Semilunar valves open

    4. Papillary muscles   Aortic semilunar valve   Tricuspid valve   Chordae tendineae

    5. Tricuspid valve   Mitral valve   Bicuspid valve   Left AV valve

    6. Foramen ovale   Ligamentum arteriosum   Coronary sinus   Fetal shunt remnants

    7. Increased heart rate   Vagus nerves   Exercise   Sympathetic nerves

    8. Ischemia   Infarct   Scar tissue repair   Heart block

Cardiovascular System: Blood Vessels

12. Complete the following statements concerning blood vessels.

    1. _____

    2. _____

    3. _____

    4. _____

    5. _____

    6. _____

    7. _____

    The central cavity of a blood vessel is called the __(1)__ . Reduction of the diameter of this cavity is called __(2)__ , and enlargement of the vessel diameter is called __(3)__ . Blood is carried to the heart by __(4)__ and away from the heart by __(5)__ . Capillary beds are supplied by __(6)__ and drained by __(7)__ .

13. Briefly explain in the space provided why valves are present in veins but not in

    arteries. _____

    _____

14. Name two events *occurring within the body* that aid in venous return. Place your responses in the blanks that follow.

    _____ and _____

15. First, select different colors for each of the three blood vessel tunics listed in the key choices and illustrated in Figure 11-4 on p. 168. Color the color coding circles and the corresponding structures in the three diagrams. In the blanks beneath the illustrations correctly identify each vessel type. In the additional spaces provided, list the structural details that allowed you to make the

identifications. Then, using the key choices, identify the blood vessel tunics described in each of the following descriptions. Insert the term or letter of the key choice in the answer blanks.

KEY CHOICES:

A. ◯ Tunica intima   **B.** ◯ Tunica media   **C.** ◯ Tunica externa

_____ **1.** Single thin layer of endothelium

_____ **2.** Bulky middle coat, containing smooth muscle and elastin

_____ **3.** Provides a smooth surface to decrease resistance to blood flow

_____ **4.** The only tunic of capillaries

_____ **5.** Also called the adventitia

_____ **6.** The only tunic that plays an active role in blood pressure regulation

_____ **7.** Supporting, protective coat

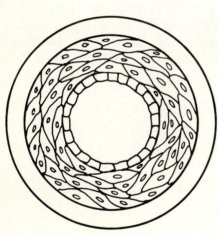
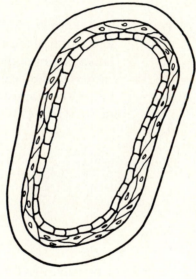

Figure 11-4

A. _____   **B.** _____   **C.** _____

_____   _____   _____

_____   _____   _____

16. Figures 11-5 and 11-6 on pp. 169 and 170 illustrate the location of the most important arteries and veins of the body. The veins are shown in Figure 11-5. Color the veins blue and then identify each vein provided with a leader line on the figure. The arteries are shown in Figure 11-6. Color them red and then identify those indicated by leader lines on the figure. NOTE: If desired, the vessels identified may be colored differently to aid you in their later identification.

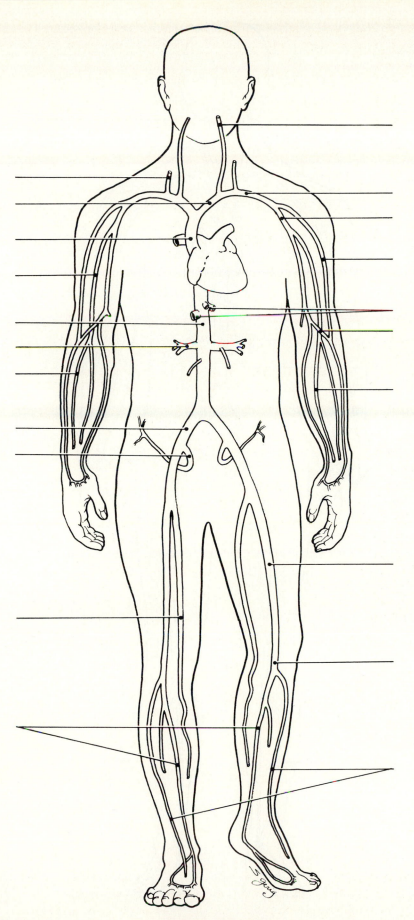

Figure 11-5: Veins

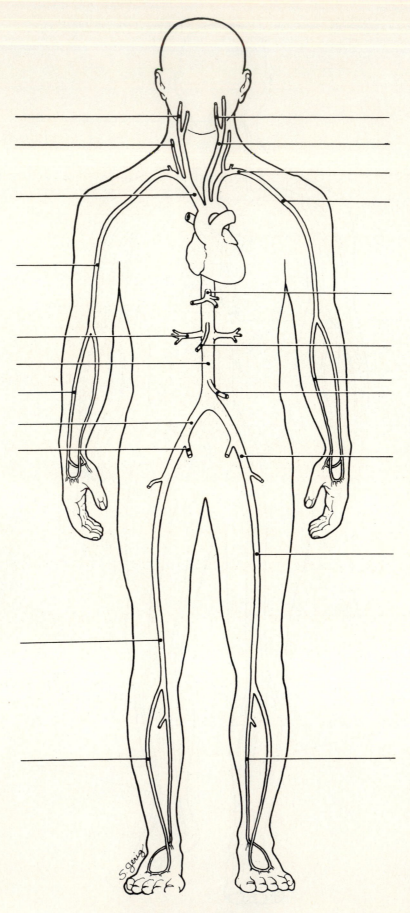

Figure 11-6: Arteries

17. Using key choices, identify the *veins* described as follows. Place the correct term or letter response in the answer blanks.

KEY CHOICES:

A. Anterior tibial	**G.** Common iliac	**M.** Hepatic portal	**S.** Radial
B. Azygos	**H.** Femoral	**N.** Inferior mesenteric	**T.** Renal
C. Basilic	**I.** Gastric	**O.** Inferior vena cava	**U.** Subclavian
D. Brachiocephalic	**J.** Gonadal	**P.** Internal iliac	**V.** Superior mesenteric
E. Cardiac	**K.** Greater saphenous	**Q.** Internal jugular	**W.** Superior vena cava
F. Cephalic	**L.** Hepatic	**R.** Posterior tibial	**X.** Ulnar

_____ **1.** _____ **2.** Deep veins, draining the forearm

_____ **3.** Vein that receives blood from the arm via the axillary vein

_____ **4.** Vein that drains the venous blood from the myocardium of the heart into the coronary sinus

_____ **5.** Vein that drains the kidney

_____ **6.** Vein that drains the dural sinuses of the brain

_____ **7.** Two veins that join to become the superior vena cava

_____ **8.** _____ **9.** Veins that drain the leg and foot

_____ **10.** Large vein that carries nutrient-rich blood from the digestive organs to the liver for processing

_____ **11.** Superficial vein that drains the lateral aspect of the arm

_____ **12.** Vein that drains the ovaries or testes

_____ **13.** Vein that drains the thorax, empties into the superior vena cava

_____ **14.** Largest vein below the thorax

_____ **15.** Vein that drains the liver

_____ **16.** _____ **17.** _____ **18.** Three veins that form/empty into the hepatic portal vein

_____ **19.** Longest superficial vein of the body; found in the leg

_____ **20.** Vein that is formed by the union of the external and internal iliac veins

_____ **21.** Deep vein of the thigh

18. Using key choices, identify the *arteries* described as follows. Place the correct term or letter response in the spaces provided.

KEY CHOICES:

A. Anterior tibial	**H.** Coronary	**O.** Intercostals	**V.** Renal
B. Aorta	**I.** Deep femoral	**P.** Internal carotid	**W.** Subclavian
C. Brachial	**J.** Dorsalis pedis	**Q.** Internal iliac	**X.** Superior mesenteric
D. Brachiocephalic	**K.** External carotid	**R.** Peroneal	**Y.** Vertebral
E. Celiac trunk	**L.** Femoral	**S.** Phrenic	**Z.** Ulnar
F. Common carotid	**M.** Hepatic	**T.** Posterior tibial	
G. Common iliac	**N.** Inferior mesenteric	**U.** Radial	

_____ **1.** _____ **2.** Two arteries formed by the division of the brachiocephalic artery

_____ **3.** First artery that branches off the ascending aorta; serves the heart

_____ **4.** _____ **5.** Two paired arteries, serving the brain

_____ **6.** Largest artery of the body

_____ **7.** Arterial network on the dorsum of the foot

_____ **8.** Artery that serves the posterior thigh

_____ **9.** Artery that supplies the diaphragm

_____ **10.** Artery that splits to form the radial and ulnar arteries

_____ **11.** Artery generally auscultated to determine blood pressure in the arm

_____ **12.** Artery that supplies the last half of the large intestine

_____ **13.** Artery that serves the pelvis

_____ **14.** External iliac becomes this artery on entering the thigh

_____ **15.** Major artery serving the arm

_____ **16.** Artery that supplies most of the small intestine

_____ **17.** The terminal branches of the dorsal, or descending, aorta

_____ **18.** Arterial trunk that has three major branches, which serve the liver, spleen, and stomach

_____ **19.** Major artery, serving the tissues external to the skull

_____ **20.** _____ **21.** _____ **22.** Three
arteries, serving the leg inferior to the knee

_____ **23.** Artery generally used to feel the pulse at the wrist

19. Figure 11-7 is a diagram of the hepatic portal circulation. Select different colors
for the structures listed below and use them to color the color coding circles
and corresponding structures on the illustration.

◯ Inferior mesenteric vein   ◯ Splenic vein   ◯ Hepatic portal vein

◯ Superior mesenteric vein   ◯ Gastric vein

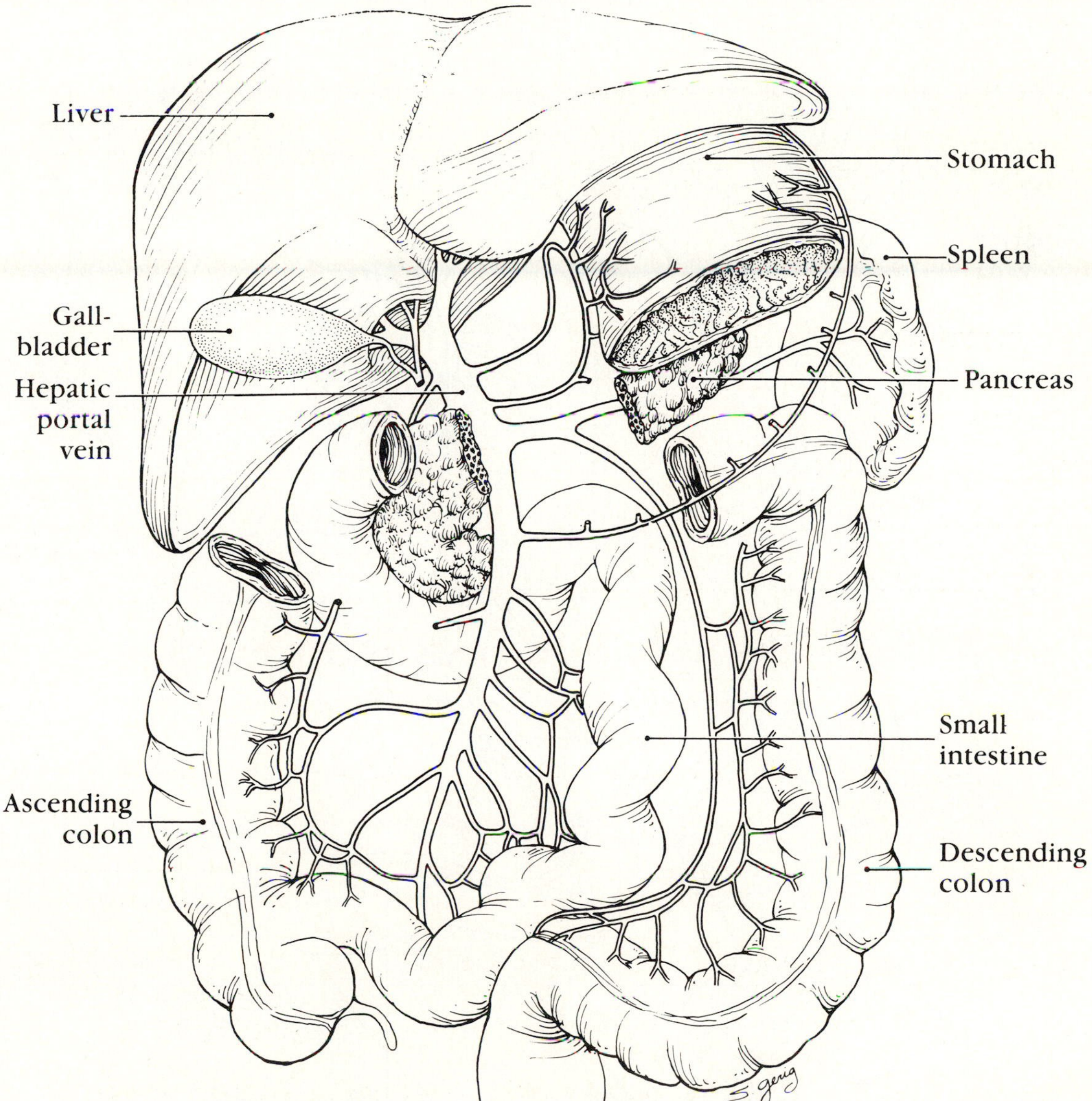

Figure 11-7

20. Figure 11-8 illustrates the special fetal structures listed below. Select different colors for each and use them to color coding circles and corresponding structures in the diagram.

◯ Foramen ovale   ◯ Ductus arteriosus   ◯ Ductus venosus

◯ Umbilical arteries   ◯ Umbilical cord   ◯ Umbilical vein

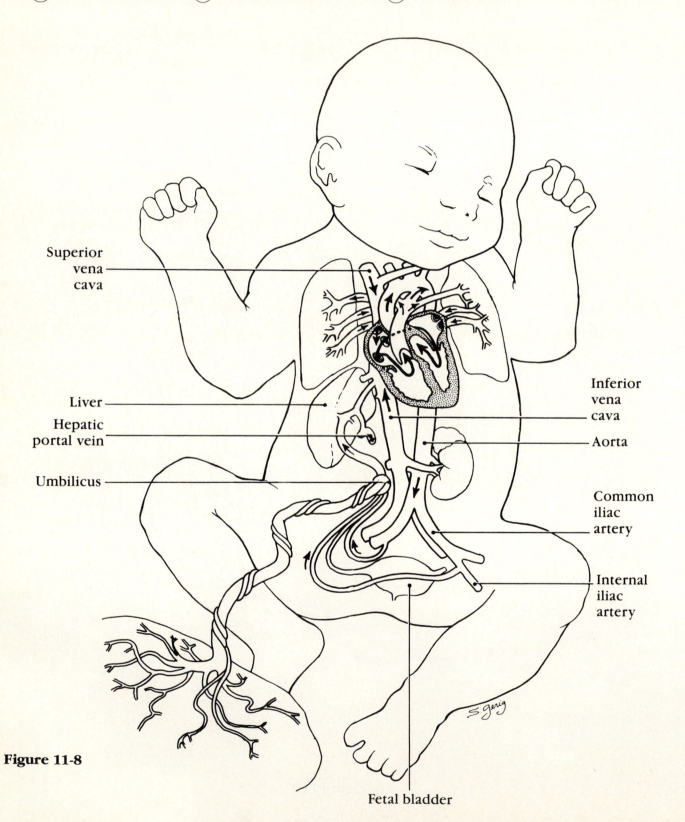

Superior
vena
cava

Inferior
vena
cava

Liver

Hepatic
portal vein

Aorta

Umbilicus

Common
iliac
artery

Internal
iliac
artery

Figure 11-8

Fetal bladder

21. Figure 11-9 illustrates the arterial circulation of the brain. Select different colors for the following structures and use them to color the coding circles and corresponding structures in the diagram.

○ Basilar artery   ○ Communicating branches

○ Anterior cerebral arteries   ○ Middle cerebral arteries

○ Posterior cerebral arteries

Assume there is a clot in the right internal carotid artery. In the space provided, briefly describe another route that might be used to provide blood to the areas served by its branches.

_____

_____

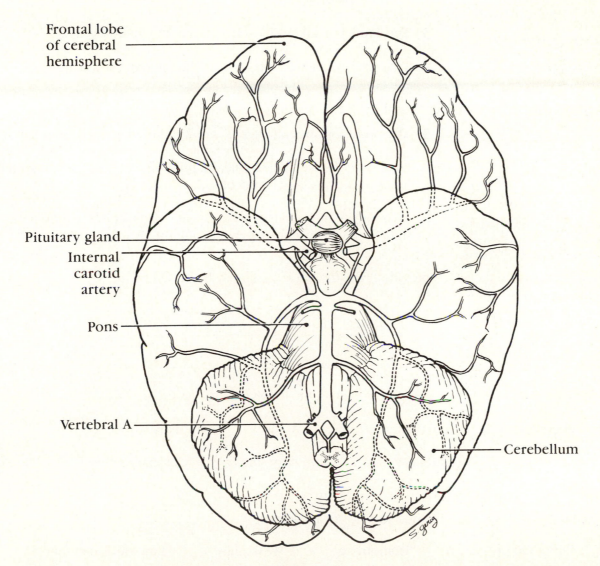

Figure 11-9

22. Eight structures unique to the special circulations of the body are described here. Identify each, using the key choices. Place the correct terms or letters in the answer blanks.

KEY CHOICES:

A. Anterior cerebral artery   **E.** Ductus venosus   **H.** Posterior cerebral artery

B. Basilar artery   **F.** Foramen ovale   **I.** Umbilical artery

C. Circle of Willis   **G.** Middle cerebral artery   **J.** Umbilical vein

D. Ductus arteriosus

_____ **1.** An anastomosis that allows communication between the posterior and anterior blood supplies of the brain

_____ **2.** The vessel carrying oxygen and nutrient-rich blood to the fetus from the placenta

_____ **3.** The shunt that allows most fetal blood to bypass the liver

_____ **4.** Two pairs of arteries, arising from the internal carotid artery

_____

_____ **5.** The posterior cerebral arteries, serving the brain, arise from here

_____ **6.** Fetal shunt between the aorta and pulmonary trunk that allows the lungs to be bypassed by the blood

_____ **7.** Opening in the interatrial septum that shunts fetal blood from the right to the left atrium, thus bypassing the fetal lungs

23. Briefly explain in the space provided why the lungs are largely bypassed by the circulating blood in the fetus.

_____

_____

24. Circle the term that does not belong in each of the following groupings.

    1. Pulmonary trunk   Lobar arteries   Aorta   Pulmonary capillaries

    2. Carotid artery   Cardiac vein   Coronary sinus   Coronary artery

    3. Increased venous return   Respiratory pump
    Milking action of skeletal muscles   Vasodilation

    4. High blood pressure   Hemorrhage   Weak pulse   Low cardiac output

    5. Resistance   Friction   Vasodilation   Vasoconstriction

    6. High pressure   Vein   Artery   Spurting blood

25. Indicate what effect the following factors have on blood pressure. Indicate an increase in pressure by *I* and a decrease in pressure by *D*. Place the correct letter response in the answer blanks.

_____ **1.** Increased diameter of the arterioles   _____ **8.** Physical exercise

_____ **2.** Increased blood viscosity   _____ **9.** Physical training

_____ **3.** Increased cardiac output   _____ **10.** Alcohol

_____ **4.** Increased pulse rate   _____ **11.** Hemorrhage

_____ **5.** Anxiety, fear   _____ **12.** Nicotine

_____ **6.** Increased urine output   _____ **13.** Arteriosclerosis

_____ **7.** Sudden change in position from reclining to standing

26. Respond to the following exercise by placing brief answers in the spaces provided. Assume someone has been injured in an automobile accident and is bleeding profusely. What pressure point could you compress to help stop the bleeding from the following areas?

_____ **1.** Thigh   _____ **4.** Lower jaw

_____ **2.** Forearm   _____ **5.** Thumb

_____ **3.** Calf

27. If the following statements are true, insert *T* in the answer blanks. If any of the statements are false, correct the underlined words by inserting the correct word in the answer blank.

_____ **1.** Renin, released by the kidneys, causes a <u>decrease</u> in blood pressure.

_____ **2.** The decreasing efficiency of the <u>sympathetic</u> nervous system vasoconstrictor functioning, due to aging, leads to a type of hypotension called <u>sympathetic</u> hypotension.

_____ **3.** Two body organs in which vasoconstriction rarely occurs are the heart and the <u>kidneys</u>.

_____ **4.** A <u>sphygmomanometer</u> is used to take the apical pulse.

_____ **5.** The pulmonary circulation is a <u>high</u> pressure circulation.

_____ **6.** The fetal equivalent of (functional) lungs and liver is the <u>placenta</u>.

_____ **7.** Cold has a <u>vasodilating</u> effect.

_____ **8.** <u>Thrombophlebitis</u> is called the silent killer.

28. The following section relates to understanding blood pressure and pulse. Match the items given in Column B with the appropriate descriptions provided in Column A. Place the correct term or letter response in the answer blanks.

	Column A	Column B
_____	1. Expansion and recoil of an artery during heart activity	**A.** Over arteries
_____	2. Pressure exerted by the blood against the blood vessel walls	**B.** Blood pressure
_____	3. _____ 4. Factors related to blood pressure	**C.** Cardiac output
		D. Constriction of arterioles
_____	5. Event primarily responsible for peripheral resistance	**E.** Diastolic blood pressure
_____	6. Blood pressure during heart contraction	**F.** Peripheral resistance
_____	7. Blood pressure during heart relaxation	**G.** Pressure points
		H. Pulse
_____	8. Site where blood pressure determinations are normally made	**I.** Sounds of Korotkoff
_____	9. Points at the body surface where the pulse may be felt.	**J.** Systolic blood pressure
_____	10. Sounds heard over a blood vessel when the vessel is partially compressed	**K.** Over veins

Lymphatic System

29. Match the terms in Column B with the appropriate descriptions in Column A. More than one choice may apply in some cases.

	Column A	Column B
_____	1. The largest lymphatic organ; a blood reservoir	**A.** Lymph nodes
_____	2. Filter lymph	**B.** Peyer's patches
_____	3. Particularly large and important during youth; produces hormones that help to program the immune system	**C.** Spleen
		D. Thymus
_____	4. Collectively called GALT	**E.** Tonsils
_____	5. Removes aged and defective red blood cells	
_____	6. Prevents bacteria from breaching the intestinal wall	
_____	7. Site of hemopoiesis in embryos	

30. Complete the following statements by inserting your responses in the answer blanks.

_____ 1.

_____ 2.

_____ 3.

_____ 4.

_____ 5.

_____ 6.

_____ 7.

_____ 8.

_____ 9.

_____ 10.

_____ 11.

_____ 12.

The lymphatic system is a specialized subdivision of the circulatory system. Although the cardiovascular system has a pump (the heart) and arteries, veins, and capillaries, the lymphatic system lacks two of these structures—the __(1)__ and __(2)__ . Like the __(3)__ of the circulatory system, the __(3)__ of the lymphatic system are also equipped with __(4)__ to prevent backflow in the vessels. The lymphatic vessels act primarily to pick up leaked fluid, now called __(5)__ , and return it to the bloodstream. The major duct receiving the drainage from the right side of the head and thorax and right arm is the __(6)__ . The rest of the body's lymphatic vessels drain into the __(7)__ . The second most abundant type of organs composing the lymphatic system are the __(8)__ . They are found throughout the body along the course of the lymphatic vessels, but there are particularly large clusters in the __(9)__ , __(10)__ , and __(11)__ regions. Basically, the lymph nodes act to __(12)__ the body by removal of bacteria (or other debris) from the lymphatic stream.

Developmental Aspects of the Circulatory System

31. Complete the following statements by inserting your responses in the answer blanks.

_____ 1.

_____ 2.

_____ 3.

_____ 4.

_____ 5.

_____ 6.

_____ 7.

_____ 8.

_____ 9.

_____ 10.

_____ 11.

The heart forms early and is acting as a functional pump by the __(1)__ week of development. The special bypass structures that exist to bypass the fetal lungs and liver become __(2)__ shortly after birth. Congenital heart defects (some resulting from the failure of the bypass structures to close) account for half of all infant __(3)__ resulting from congenital defects.

Regular __(4)__ increases the efficiency of the cardiovascular system and helps to slow the progress of __(5)__ . A vascular problem that affects many in "standing professions" is __(6)__ . In this condition, the valves become incompetent, and the veins become twisted and enlarged, particularly in the __(7)__ and __(8)__ . Very few escape the gradual development of arteriosclerosis, which eventually leads to __(9)__ and __(10)__ . These conditions, in turn, often result in myocardial infarcts, or __(11)__ . Risk factors for both conditions include __(12)__ , __(13)__ , __(14)__ , and __(15)__ .

_____ 12.

_____ 13.

_____ 14.

_____ 15.

_____ 16.

_____ 17.

Lymphatic vessels derive from the "budding" of __(16)__.
Other than the __(17)__, most lymphoid organs are poorly
developed before birth.

The Incredible Journey:
A Visualization Exercise
for the Circulatory System

All about you are huge white cords, hanging
limply from two flaps of endothelial tissue. . .

32. Where necessary, complete the statements by inserting the missing word(s) in
the answer blanks.

_____ 1.

_____ 2.

_____ 3.

_____ 4.

_____ 5.

_____ 6.

_____ 7.

_____ 8.

_____ 9.

_____ 10.

_____ 11.

_____ 12.

_____ 13.

Your journey starts in the pulmonary vein and includes a trip
to part of the systemic circulation and a special circulation.
You ready your equipment and prepare to be miniaturized
and injected into your host.

Almost immediately after injection, you find yourself swept
into a good-sized chamber, the __(1)__. However, you do not
stop in this chamber, but continue to plunge downward into
a larger chamber below. You land with a large splash and
examine your surroundings. All about you are huge white
cords, hanging limply from two flaps of endothelial tissue
far above you. You report that you are sitting in the __(2)__
chamber of the heart, seeing the flaps of the __(3)__ valve
above you. The valve is open and its anchoring cords,
the __(4)__, are lax. Since this valve is open, you conclude
that the heart is in the __(5)__ phase of the cardiac cycle.

Gradually you notice that the chamber walls seem to be
closing in. You hear a thundering boom, and the whole
chamber vibrates as the valve slams shut above you. The
cords, now rigid and strained, form a cage about you, and
you feel extreme external pressure. Obviously, the heart is
in a full-fledged __(6)__. Then, high above on the right, the
"roof" opens, and you are forced through this __(7)__ valve.
A fraction of a second later, you hear another tremendous

_____ 14.

_____ 15.

_____ 16.

_____ 17.

_____ 18.

_____ 19.

_____ 20.

boom that sends shock waves through the whole area. Out of the corner of your eye, you see that the valve below you is closed, and it looks rather like a pie cut into three wedges.

As you are swept along in this huge artery, the __(8)__ , you pass several branch-off points, but continue to career along, straight down at a dizzying speed until you approach the __(9)__ artery, feeding the small intestine. After entering this artery and passing through successively smaller and smaller subdivisions of it, you finally reach the capillary bed of the small intestine. You watch with fascination as nutrient molecules move into the blood through the single layer of __(10)__ cells forming the capillary wall. As you move to the opposite shore of the capillary bed, you enter a venule and begin to move superiorly once again. The venules draining the small intestine combine to form the __(11)__ vein, which in turn combines with the __(12)__ vein to form the hepatic portal vein that carries you into the liver. As you enter the liver, you are amazed at the activity there. Six-sided hepatic cells, responsible for storing glucose and making blood proteins, are literally grabbing __(13)__ out of the blood as it percolates slowly past them. Protective __(14)__ cells are removing bacteria from the slowly moving blood. Leaving the liver through the __(15)__ vein, you almost immediately enter the huge __(16)__ , which returns blood from the lower part of the body to the __(17)__ of the heart. From here, you move consecutively through the right chambers of the heart into the __(18)__ artery, which carries you to the capillary beds of the __(19)__ and then back to the left side of the heart once again. After traveling through the left side of the heart again, you leave your host when you are aspirated out of the __(20)__ artery, which extends from the aorta to the axillary artery of the armpit.

Body Defenses

The immune system is a unique functional system made up of billions of individual cells, the bulk of which are lymphocytes. The sole function of this highly specialized defensive system is to protect the body against an incredible array of pathogens. In general, these "enemies" fall into three major camps: (1) microorganisms (bacteria, viruses, and fungi) that have gained entry into the body, (2) foreign tissue cells that have been transplanted (or, in the case of red blood cells, infused) into the body, and (3) the body's own cells that have become cancerous. The result of the immune system's activities is immunity, or specific resistance to disease.

The body is also protected by a number of nonspecific defenses provided by intact surface membranes such as skin and mucosae, and a variety of cells and chemicals can quickly mount the attack against foreign substances. The specific and nonspecific defenses enhance each other's effectiveness.

This chapter first focuses on the nonspecific body defenses. It then follows the development of the immune system and describes the mode of functioning of its mediating cells, lymphocytes and macrophages. The chemicals that act to mediate or amplify the immune response are also discussed.

Nonspecific Body Defenses

1. Indicate the sites of activity or the secretions of the nonspecific defenses by writing the correct terms in the answer blanks.

    1. Lysozyme is found in the body secretions called _____ and _____ .

    2. Fluids with an acid pH are found in the _____ and _____ .

    3. Sebum is a product of the _____ glands and acts at the surface of the

    _____ .

    4. Mucus is produced by mucus-secreting glands found in the respiratory and _____ system mucosae.

2.   Match the terms in Column B with the descriptions of the nonspecific defenses of the body in Column A. More than one choice may apply.

Column A

_____ **1.** Have antimicrobial activity

_____ **2.** Provide mechanical barriers

_____ **3.** Provide chemical barriers

_____ **4.** Entraps microorganisms entering the respiratory passages

_____ **5.** Part of the first line of defense

Column B

A. Acids

B. Lysozyme

C. Mucosae

D. Mucus

E. Protein-digesting enzymes

F. Sebum

G. Skin

3.   Describe the protective role of cilia. _____

_____

_____

4.   Define *phagocytosis*. _____

_____

_____

5.   Circle the term that does not belong in each of the following groupings.

   1. Redness   Pain   Swelling   Itching   Heat

   2. Neutrophils   Macrophages   Phagocytes   Natural killer cells

   3. Inflammatory chemicals   Histamine   Kinins   Complement proteins
   Interferon

   4. Intact skin   Intact mucosae   Lysozyme   Inflammation
   First line of defense

   5. Interferons   Anti-viral   Anti-bacterial   Anti-cancer

6.   List the three important accomplishments of the inflammatory response.

    1. _____

    2. _____

    3. _____

7.   Match the terms in Column B with the descriptions in Column A concerning events of the inflammatory response.

Column A

_____ **1.** Accounts for redness and heat in an inflamed area

_____ **2.** Inflammatory chemical released by degranulating mast cells

_____ **3.** Promote release of white blood cells from the bone marrow

_____ **4.** Cellular migration directed by a chemical gradient

_____ **5.** Results from accumulation of fluid leaked from the bloodstream

_____ **6.** Phagocytic progeny of monocytes

_____ **7.** Leukocytes pass through the wall of a capillary

_____ **8.** First phagocytes to migrate into the injured area

_____ **9.** Walls off the area of injury

Column B

A. Chemotaxis

B. Diapedesis

C. Edema

D. Fibrin mesh

E. Histamine

F. Increased blood flow to an area

G. Inflammatory chemicals (including E)

H. Macrophages

I. Neutrophils

8.   Complete the following description of the activation and activity of complement by writing the missing terms in the answer blanks.

_____ **1.**

_____ **2.**

_____ **3.**

_____ **4.**

_____ **5.**

_____ **6.**

Complement is a system of plasma __(1)__ that circulate in the blood in an inactive form. Complement is __(2)__ when it becomes attached to the surface of foreign cells (bacteria, fungi, red blood cells). The result of this complement fixation is that some of the complement elements, called the __(3)__, become inserted in the membrane of the foreign cell. This forms a __(4)__, which causes __(5)__ of the foreign cell. Some of the chemicals released during complement fixation enhance phagocytosis. This is called __(6)__. Others amplify the inflammatory response.

9. Describe the event that leads to the synthesis of interferon and the result of its synthesis.

_____

_____

Specific Body Defenses: The Immune System

Cells and Tissues of the Immune Response

10. Using the key choices, select the term that correctly completes each statement. Insert the appropriate term or letter in the answer blanks.

KEY CHOICES:

A. Antigen(s)   **D.** Cellular immunity   **G.** Lymph nodes

B. B cells   **E.** Humoral immunity   **H.** Macrophages

C. Blood   **F.** Lymph   **I.** T cells

_____ 1.

_____ 2.

_____ 3.

_____ 4.

_____ 5.

_____ 6.

_____ 7.

_____ 8.

_____ 9.

Immunity is resistance to disease resulting from the presence of foreign substances or __(1)__ in the body. When this resistance is provided by antibodies released to body fluids, the immunity is called __(2)__. When living cells provide the protection, the immunity is referred to as __(3)__. The major actors in the immune response are two lymphocyte populations, the __(4)__ and the __(5)__. Phagocytic cells that act as accessory cells in the immune response are the __(6)__. Because pathogens are likely to use both __(7)__ and __(8)__ as a means of getting around the body, __(9)__ and other lymphatic tissues (which house the immune cells) are in an excellent position to detect their presence.

11. A schematic of the life cycle of the lymphocytes involved in immunity is shown in Figure 12-1. First, select different colors for the areas listed below and use them to color the coding circles and the corresponding regions in the figure. If there is overlap, use stripes of a second color to indicate the second identification. Then respond to the statements following the figure, which relate to the two-phase differentiation process of B and T cells.

◯ Area where immature lymphocytes arise

◯ Area seeded by immunocompetent B and T cells

◯ Area where T cells become immunocompetent

◯ Area where the antigen challenge and clonal selection are likely to occur

◯ Area where B cells become immunocompetent

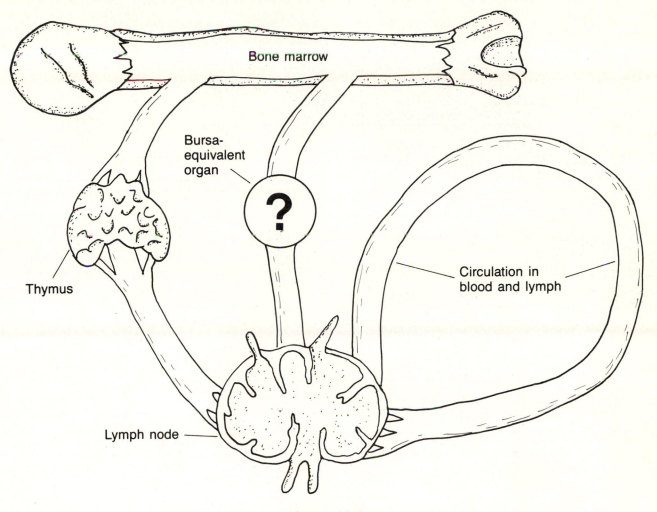

Figure 12-1

1. What signifies that a lymphocyte has become immunocompetent?

_____

2. During what period of life does the development of immunocompetence occur?

_____

3. What determines which antigen a particular T or B cell will be able to recognize? (circle one)

  its genes   "its" antigen

4. What triggers the process of clonal selection in a T or B cell? (circle one)

  its genes   binding to "its" antigen

5. Which organ is believed to be the bursa-equivalent organ?

_____

6. During the development of immunocompetence (during which B or T cells develop the ability to recognize a particular antigen) the ability to tolerate _____ must also occur if the immune system is to function normally.

12. T cells and B cells exhibit certain similarities and differences. Check (✓) the appropriate spaces in the table below to indicate the lymphocyte type that exhibits each characteristic.

Characteristic	T cell	B cell
Originates in bone marrow from stem cells called hemocytoblasts		
Progeny are plasma cells		
Progeny include supressors, helpers, and killers		
Progeny include memory cells		
Is responsible for directly attacking foreign cells or viruses that enter the body		
Produces antibodies that are released to body fluids		
Bears a cell-surface receptor capable of recognizing a specific antigen		
Forms clones upon stimulation		
Accounts for most of the lymphocytes in the circulation		

13. Complete the following statements relating to antigens by writing the missing terms in the answer blanks.

_____ 1.

_____ 2.

_____ 3.

_____ 4.

Antigens are substances capable of mobilizing the __(1)__. Of all the foreign molecules that act as complete antigens, __(2)__ are the most potent. Small molecules are not usually antigenic, but when they bind to self cell surface proteins they may act as __(3)__, and then the complex is recognized as foreign, or __(4)__.

14. Several populations of T cells exist. Match the terms in Column B to the descriptions in Column A. Place the correct term or letter response in the answer blanks.

Column A

_____ 1. Binds with B cells and releases chemicals that activate B cells

_____ 2. Releases chemicals that activate killer T cells, macrophages, and a variety of nonimmune cells, but must be activated itself by recognizing both its antigen and a self-protein presented on the surface of a macrophage

_____ 3. Turns off the immune response when the "enemy" has been routed

_____ 4. Directly attacks, and causes lysis of, cellular pathogens

Column B

A. Helper T cell

B. Killer T cell

C. Suppressor T cell

15. Using the key choices, select the terms that correspond to the descriptions of substances or events by inserting the appropriate term or letter in the answer blanks.

KEY CHOICES:

A. Anaphylaxis   D. Complement   G. Lymphokine

B. Antibodies   E. Inflammation   H. Macrophage activating factor

C. Chemotaxis factors   F. Interferon   I. Monokines

_____ 1. A protein released by macrophages and activated T cells that helps to protect other body cells from viral multiplication

_____ 2. Any types of molecules that attract neutrophils and other protective cells into a region where an immune response is ongoing

_____ 3. Proteins released by plasma cells that mark antigens for destruction by phagocytes or complement

_____ 4. A consequence of the release of histamine from mast cells and of complement activation

_____ 5. C, F, and H are examples of this class of molecules

_____ 6. A group of plasma proteins that amplifies the immune response by causing lysis of cellular pathogens once it has been "fixed" to their surface.

_____ 7. Class of chemicals released by macrophages; interleukin 1 and interferon are examples.

16. Figure 12-2 is a flowchart of the immune response that tests your understanding of the interrelationships of that process. Several terms have been omitted from this schematic. First, complete the figure by inserting appropriate terms from the key choices below. (Note that oval blanks indicate that the required term identifies a cell type and rectangular blanks represent the names of chemical molecules. Also note that solid lines represent stimulatory or enhancing effects whereas broken lines indicate inhibition.) Then color the coding circles and the corresponding ovals, indicating the cell types identified.

KEY CHOICES:

Cell types: Molecules:

◯ B cell   Antibodies

◯ Helper T cell   Chemotaxis factors

◯ Killer T cell   Complement

◯ Macrophage   Interferon

◯ Memory B cell   Interleukin

◯ Memory T cell   Lymphokines

◯ Neutrophils   Lymphotoxins

◯ Plasma cell   Macrophage activating factor

◯ Suppressor T cell   Suppressor factors

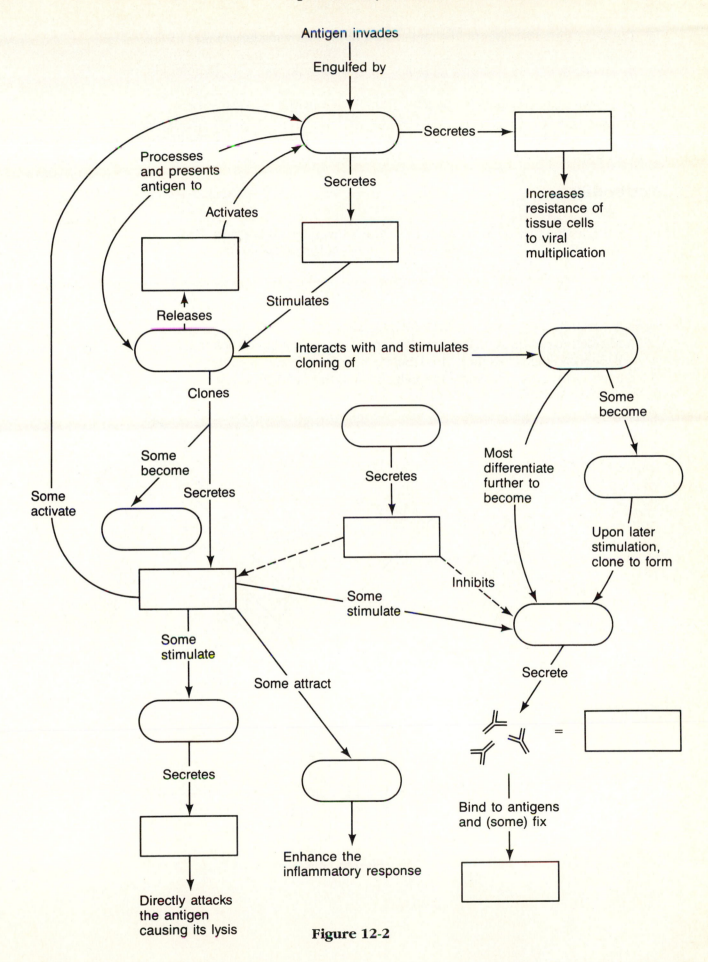

Figure 12-2

17. Circle the term that does not belong in each of the following groupings.

    1. Antibodies   Gamma globulin   Lymphokines   Immunoglobulins

    2. Protein   Complete antigen   Nucleic acid   Hapten

    3. Lymph nodes   Liver   Spleen   Thymus   Bone marrow

Antibodies

18. The basic structure of an antibody molecule is diagrammed in Figure 12-3. Select different colors, and color in the coding circles below and the corresponding areas on the diagram.

          ◯ heavy chains   ◯ light chains

Add labels to the diagram to correctly identify the type of bonds holding the polypeptide chains together. Also label the constant (C) and variable (V) regions of the antibody, and add "polka dots" to the variable portions.

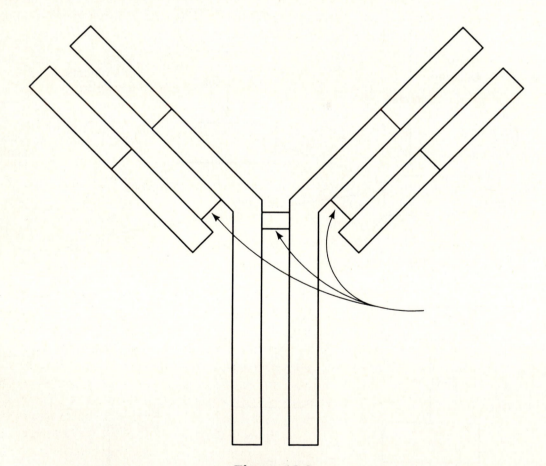

Figure 12-3

1. Which portion of the antibody—V or C—is its antigen-binding site?

_____

2. Which portion acts to determine antibody class and specific function?

_____

19. Match the antibody classes in Column B to their descriptions in Column A.
Place the correct term or letter response in the answer blanks.

Column A	Column B
_____ **1.** Bound to the surface of a B cell	**A.** IgA
_____ **2.** Crosses the placenta	**B.** IgD
_____ **3.** The first antibody released during the primary response	**C.** IgE
_____ **4.** Fixes complement (two classes)	**D.** IgG
_____ **5.** Is a pentamer	**E.** IgM
_____ **6.** The most abundant antibody found in blood plasma and the chief antibody released during secondary responses	
_____ **7.** Binds to the surface of mast cells and mediates an allergic response	
_____ **8.** Predominant antibody found in mucus, saliva, and tears	

20. Complete the following descriptions of antibody function by writing the missing terms in the answer blanks.

_____ **1.**

_____ **2.**

_____ **3.**

_____ **4.**

_____ **5.**

_____ **6.**

_____ **7.**

Antibodies can inactivate antigens in various ways, depending on the nature of the __(1)__. __(2)__ is the chief ammunition used against cellular antigens such as bacteria and mismatched red blood cells. The binding of antibodies to sites on bacterial exotoxins or viruses that can cause cell injury is called __(3)__. The cross linking of cellular antigens into large latices by antibodies is called __(4)__; Ig __(5)__, with its 10 antigen binding sites, is particularly efficient in this mechanism. When molecules are cross-linked into latices by antibodies, the mechanism is more properly called __(6)__. In virtually all these cases, the protective mechanism mounted by the antibodies serves to disarm and/or immobilize the antigens until they can be disposed of by __(7)__.

21. Determine whether each of the following situations provides, or is an example of, active or passive immunity. If passive, insert a *P* in the blank; if active, insert an *A* in the blank.

    _____ **1.** An individual receives Sabin polio vaccine

    _____ **2.** Antibodies migrate through a pregnant woman's placenta into the vascular system of her fetus

    _____ **3.** A student nurse receives an injection of gamma globulin (containing antibodies to the hepatitis virus) after she has been exposed to viral hepatitis

    _____ **4.** "Borrowed" immunity

    _____ **5.** Immunologic memory is provided

    _____ **6.** An individual suffers through chicken pox

22. There are several important differences between primary and secondary immune response(s) to antigens. If the following statements best describe a primary response, insert a *P* in the blank; if a secondary response, insert an *S* in the blank.

    _____ **1.** The initial response to an antigen; gearing-up stage

    _____ **2.** A lag period of several days occurs before antibodies specific to the antigen appear in the bloodstream

    _____ **3.** Antibody levels increase rapidly and remain high for an extended period

    _____ **4.** Immunologic memory is established

    _____ **5.** The second, third, and subsequent responses to the same antigen

Disorders of Immunity

23. Using the key choices, identify the type of immunity disorder described. Insert the appropriate term or letter in the answer blank.

    KEY CHOICES:

    A. Allergy   **B.** Autoimmune disease   **C.** Immunodeficiency

    _____ **1.** AIDS and congenital thymic aplasia

    _____ **2.** The immune system mounts an extraordinarily vigorous response to an otherwise harmless antigen

    _____ **3.** A hypersensitivity reaction

_____   **4.** Occurs when the production or activity of immune cells or complement is abnormal

_____   **5.** The body's own immune system produces the disorder; a breakdown of self-tolerance

_____   **6.** Affected individuals are unable to combat infections that would present no problem for normally healthy people

_____   **7.** Multiple sclerosis and rheumatic fever

_____   **8.** Hayfever and contact dermatitis

_____   **9.** Typical symptoms of the acute response are tearing, a runny nose, and itching skin

Developmental Aspects of Body Defenses

24. Complete the following statements concerning the development and operation of the immune system during the life span by inserting your answers in the answer blanks.

_____   **1.**

_____   **2.**

_____   **3.**

_____   **4.**

_____   **5.**

_____   **6.**

_____   **7.**

_____   **8.**

_____   **9.**

_____   **10.**

_____   **11.**

_____   **12.**

The earliest lymphocyte stem cells that can be identified appear during the first month of development in the fetal __(1)__. Shortly thereafter, bone marrow becomes the lymphocyte origin site; but after birth, lymphocyte proliferation occurs in the __(2)__. The __(3)__ is the first lymphoid organ to appear, and the development of the other lymphoid organs is believed to be controlled by the thymic hormone __(4)__. The development of immunocompetence has usually been accomplished by __(5)__.

Individuals under severe stress exhibit a __(6)__ immune response, an indication that the __(7)__ system plays a role in immunity. During old age, the effectiveness of the immune system __(8)__; and elders are more at risk for __(9)__, __(10)__, and __(11)__. Part of the declining defenses may reflect the fact that __(12)__ antibodies are unable to get to the mucosal surfaces where they carry out their normal protective role.

The Incredible Journey:
A Visualization Exercise
for the Immune System

Something quite enormous and looking much like an octopus is nearly blocking the narrow tunnel just ahead.

25. Where necessary, complete statements by inserting the missing word(s) in the answer blanks.

_____ 1.

_____ 2.

_____ 3.

_____ 4.

_____ 5.

_____ 6.

_____ 7.

_____ 8.

_____ 9.

_____ 10.

_____ 11.

For this journey, you are equipped with scuba gear before you are miniaturized and injected into one of your host's lymphatic vessels. He has been suffering with a red, raw "strep throat" and has swollen cervical lymph nodes. Your assignment is to travel into a cervical lymph node and observe the activities going on there that reveal that your host's immune system is doing its best to combat the infection.

On injection, you enter the lymph with a "WHOOSH" and then bob gently in the warm yellow fluid. As you travel along, you see what seem to be thousands of spherical bacteria and a few large globular __(1)__ molecules that, no doubt, have been picked up by the tiny lymphatic capillaries. Shortly thereafter, a large dark mass looms just ahead. This has to be a __(2)__, you conclude, and you dig in your wet suit pocket to find the waterproof pen and recording tablet.

As you enter the gloomy mass, the lymphatic stream becomes shallow and begins to flow sluggishly. So that you can explore this little organ fully, you haul yourself to your feet and begin to wade through the slowly moving stream. On each bank you see a huge ball of cells that have large nuclei and such a scant amount of cytoplasm that you can barely make it out. You write, "Sighted the spherical germinal centers composed of __(3)__." As you again study one of the cell masses, you spot one cell that looks quite different and reminds you of a nest of angry hornets, because it is furiously spewing out what seems to be a horde of tiny Y-shaped "bees." "Ah ha," you think, "another valuable piece of information." You record, "Spotted a __(4)__ making and releasing __(5)__."

That done, you turn your attention to scanning the rest of the landscape. Suddenly you let out an involuntary yelp. Something quite enormous and looking much like an octopus is nearly blocking the narrow tunnel just ahead. Your mind whirls as it tries to figure out the nature of this cellular "beast" that appears to be guarding the channel. Then it hits you—this has to be a __(6)__ on the alert for foreign invaders (more properly called __(7)__), which it "eats" when it catches them. The giant cell roars, "Halt, stranger, and be recognized," and you dig frantically in your pocket for your identification pass. As you drift toward the huge cell, you hold the pass in front of you, hands trembling because you know this cell could liquify you quick as the blink of an eye. Again the cell bellows at you, "Is this some kind of a security check? I'm on the job, as you can see!" Frantically you shake your head "NO," and the cell lifts one long tentacle and allows you to pass. As you squeeze by, the cell says, "Being inside, I've never seen my body's outside. I must say, humans are a rather strange-looking lot!" Still shaking, you decide that you are in no mood for a chat and hurry along to put some distance between yourself and this guard cell.

Immediately ahead are what appear to be hundreds of the same type of cell sitting on every ledge and in every nook and cranny. Some are busily snagging and engulfing unfortunate strep bacteria that float too close. The slurping sound is nearly deafening. Then something grabs your attention: the surface of one of these cells is becoming dotted with some of the same donut-shaped chemicals that you see on the strep bacteria membranes; and a round cell, similar, but not identical, to those you earlier saw in the germinal centers, is starting to bind to one of these "doorknobs." You smile smugly, because you know you have properly identified the octopuslike cells. You then record your observations as follows: "Cells like the giant cell just identified act as __(8)__. I have just observed one in this role during its interaction with a helper __(9)__ cell."

You decide to linger a bit to see if the round cell becomes activated. You lean against the tunnel walls and watch quietly, but your wait is brief. Within minutes, the cell that was binding to the octopuslike cell begins to divide, and then its daughter cells divide again and again at a head-spinning pace. You write, "I have just witnessed the formation of a __(10)__ of like cells." Most of the daughter cells enter the lymph stream, but a few of them settle back and seem to go into a light sleep. You decide that the "napping cells" don't have any role to play in helping get rid of your host's present strep infection, but instead will provide for __(11)__ and become active at a later date.

You glance at your watch and wince as you realize that it is already 5 minutes past the time for your retrieval. You have already concluded that this is a dangerous place for those who don't "belong" and are far from sure about how long your pass is good, so you swim hurriedly from the organ into the lymphatic stream to reach your pickup spot.

Respiratory System

Body cells require an abundant and continuous supply of oxygen to carry out their activities. As cells use oxygen, they release carbon dioxide, a waste product of which the body must rid itself. The circulatory and respiratory systems are intimately involved in obtaining and delivering oxygen to body cells and in eliminating carbon dioxide from the body. The respiratory system structures are responsible for gas exchange between the blood and the external environment (that is, external respiration). The respiratory system also plays an important role in maintaining the acid–base balance of the blood.

Questions and activities in this chapter consider both the anatomy and physiology of the respiratory system structures.

Anatomy

1. The following questions refer to the primary bronchi. In the spaces provided, insert the letter *R* to indicate the right primary bronchus and the letter *L* to indicate the left primary bronchus.

   1. Which of the primary bronchi is larger in diameter? _____

   2. Which of the primary bronchi is more horizontal? _____

   3. Which of the primary bronchi is the most common site for lodging of a foreign object that has

   entered the respiratory passageways? _____

2. Complete the following statements by inserting your answers in the answer blanks.

_____ 1.

_____ 2.

_____ 3.

_____ 4.

_____ 5.

_____ 6.

_____ 7.

_____ 8.

_____ 9.

_____ 10.

_____ 11.

_____ 12.

_____ 13.

_____ 14.

_____ 15.

_____ 16.

_____ 17.

Air enters the nasal cavities of the respiratory system through the __(1)__ . The nasal cavities are divided by the midline __(2)__ . The nasal cavity mucosa has several functions. Its major functions are to __(3)__ , __(4)__ , and __(5)__ the incoming air. Mucous membrane–lined cavities called __(6)__ are found in several bones surrounding the nasal cavities. They make the skull less heavy and probably act as resonance chambers for __(7)__ . The passageway common to the digestive and respiratory systems, the __(8)__ , is often referred to as the throat; it connects the nasal cavities with the __(9)__ below. Clusters of lymphatic tissue, __(10)__ , are part of the defensive system of the body. Reinforcement of the trachea with __(11)__ rings prevents its collapse during __(12)__ changes with breathing. The fact that the rings are incomplete posteriorly allows a food bolus to bulge __(13)__ during its transport to the stomach. The larynx or voice box is built from many cartilages, but the largest are the signet ring–shaped __(14)__ and the "Adam's apple," or __(15)__ , cartilage. Within the larynx are the __(16)__ , which vibrate with exhaled air and allow an individual to __(17)__ .

3. Circle the term that does not belong in each of the following groupings.

    1. Sphenoidal   Maxillary   Mandibular   Ethmoid   Frontal

    2. Nasal cavity   Trachea   Alveolus   Bronchiole   Bronchus

    3. Apex   Base   Hilus   Larynx   Pleura

    4. Sinusitis   Peritonitis   Pleurisy   Tonsillitis   Laryngitis

    5. Laryngopharynx   Oropharynx   Transports air and food   Nasopharynx

4. Figure 13-1 is a sagittal view of the upper respiratory structures. First, correctly identify all structures provided with leader lines on the figure. Then select different colors for the structures listed below and use them to color in the coding circles and the corresponding structures on the figure

    ◯ Nasal cavity   ◯ Larynx

    ◯ Pharynx   ◯ Paranasal sinuses

    ◯ Trachea

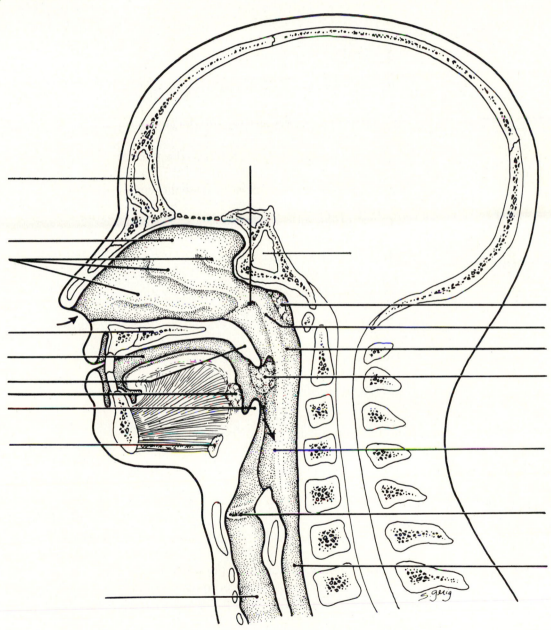

Figure 13-1

5. Using the key choices, select the terms identified in the following descriptions by inserting the appropriate term or letter in the answer blanks.

    KEY CHOICES:

    A. Alveoli   **E.** Esophagus   **H.** Parietal pleura   **K.** Trachea

    B. Bronchioles   **F.** Glottis   **I.** Phrenic nerve   **L.** Vagus nerve

    C. Conchae   **G.** Palate   **J.** Primary bronchi   **M.** Visceral pleura

    D. Epiglottis

    _____ **1.** Smallest respiratory passageways

    _____ **2.** Separates the oral and nasal cavities

    _____ **3.** Major nerve, stimulating the diaphragm

    _____ **4.** Food passageway posterior to the trachea

    _____ **5.** Closes off the larynx during swallowing

    _____ **6.** Windpipe

    _____ **7.** Actual site of gas exchanges

    _____ **8.** Pleural layer covering the thorax walls

    _____ **9.** Autonomic nervous system nerve, serving the thorax

    _____ **10.** Lumen of larynx

    _____ **11.** Fleshy lobes in the nasal cavity, which increase its surface area

6. Complete the following paragraph on the alveolar cells and their roles by writing the missing terms in the answer blanks.

    _____ **1.**

    _____ **2.**

    _____ **3.**

    _____ **4.**

    With the exception of the stroma of the lungs, which is __(1)__ tissue, the lungs are mostly air spaces, of which the alveoli comprise the greatest part. The bulk of the alveolar walls are made up of squamous epithelial cells, which are well suited for their __(2)__ function. Much less numerous cuboidal cells produce a fluid that coats the air-exposed surface of the alveolus and contains a lipid-based molecule called __(3)__ that functions to __(4)__ of the alveolar fluid.

7. Figure 13-2 is a diagram of the larynx and associated structures. On the figure, identify each of the structures listed below. Select a different color for each and use it to color in the coding circles and the corresponding structures on the figure. Then answer the questions following the diagram.

◯ Hyoid bone   ◯ Tracheal cartilages

◯ Cricoid cartilage   ◯ Epiglottis

◯ Thyroid cartilage

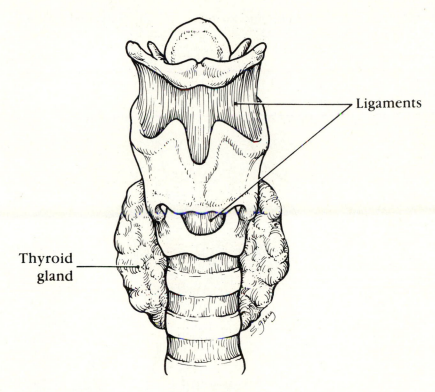

Figure 13-2

1. What are the three functions of the larynx? _____

_____

_____

2. What type of cartilage forms the epiglottis? _____

3. What type of cartilage forms the other eight laryngeal cartilages? _____

4. Explain this difference. _____

_____

_____

8. Figure 13-3 illustrates the gross anatomy of the lower respiratory system. Intact structures are shown on the left; respiratory passages are shown on the right. Select a different color for each of the structures listed below and use it to color in the coding circles and the corresponding structures on the figure. Then complete the figure by labeling the areas/structures that are provided with leader lines on the figure. Be sure to include the following: pleural space, mediastinum, apex of right lung, diaphragm, clavicle, and the base of the right lung.

⭕ Trachea   ⭕ Primary bronchi   ⭕ Visceral pleura

⭕ Larynx   ⭕ Secondary bronchi   ⭕ Parietal pleura

⭕ Intact lung

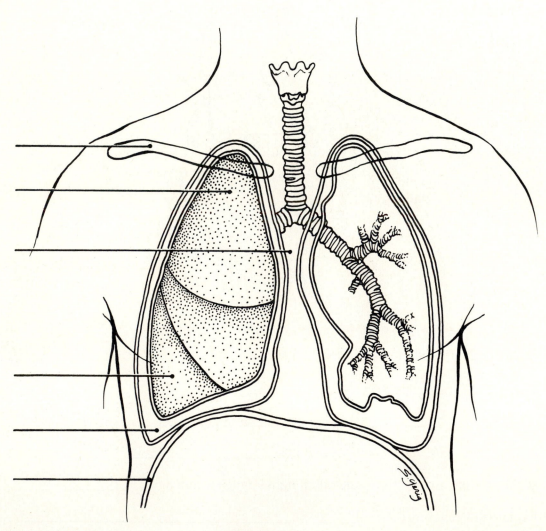

Figure 13-3

9. Figure 13-4 illustrates the microscopic structure of the respiratory unit of lung tissue. The external anatomy is shown on Figure 13-4, A. Color the intact alveoli yellow, the pulmonary capillaries red, and the respiratory bronchioles green. A cross section through an alveolus is shown on Figure 13-4, B. On this illustration color the alveolar epithelium yellow, the capillary epithelium pink, and the red blood cells in the capillary red. Also, label the alveolar chamber and color it pale blue. Finally, add the symbols for oxygen gas (O_2) and carbon dioxide gas (CO_2) in the sites where they would be in higher concentration and add arrows showing their direction of movement through the respiratory membrane.

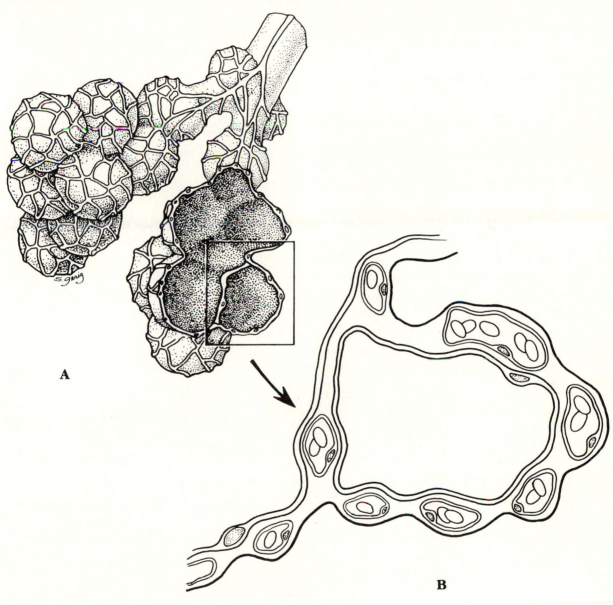

A

B

Figure 13-4

Physiology

10. Using the key choices, select the terms identified in the following descriptions
    by inserting the appropriate term or letter in the answer blanks.

    KEY CHOICES:

    A. Atmospheric pressure   **C.** Intrapleural pressure

    B. Intrapulmonary pressure   **D.** Both intrapulmonary and intrapleural pressure

    _____ **1.** Barring pneumothorax, it is always lower than atmospheric
    pressure (that is, it is negative pressure)

    _____ **2.** Pressure of air outside the body

    _____ **3.** As it decreases, air flows into the passageways of the lungs

    _____ **4.** As it increases over atmospheric pressure, air flows out of the lungs

    _____ **5.** If this pressure becomes equal to the atmospheric pressure, the
    lungs collapse

    _____ **6.** Rises well over atmospheric pressure during a forceful cough

11. Many exchanges occur within the lungs as the diaphragm (and external
    intercostal muscles) contract and then relax. These changes lead to the flow of
    air into and out of the lungs. The activity of the diaphragm is given in the left
    column of the following table. Several changes in condition are listed in the
    column heads to the right. Complete the table by checking (✔) the appropriate
    column to correctly identify the change that would be occurring relative to
    diaphragm activity in each case.

Activity of diaphragm	Changes in							
	Internal volume of thorax		Internal pressure in thorax		Size of lungs		Direction of air flow	
(↑ = increased) (↓ = decreased)	↑	↓	↑	↓	↑	↓	Into lung	Out of lung
Contracted, moves downward								
Relaxed, moves superiorly								

12. Use the key choices to respond to the following descriptions. Insert the correct
    term or letter in the answer blanks.

    KEY CHOICES:

    A. External respiration   **C.** Inspiration   **E.** Ventilation (breathing)

    B. Expiration   **D.** Internal respiration

    _____ **1.** Period of breathing when air enters the lungs

    _____ **2.** Exchange of gases between the systemic capillary blood and body
                               cells

    _____ **3.** Alternate flushing of air into and out of the lungs

    _____ **4.** Exchange of gases between alveolar air and pulmonary capillary
                               blood

13. During forced inspiration, accessory muscles are activated to raise the rib cage
    more vigorously than occurs during quiet inspiration.

    1. Name two such muscles. _____

    2. Although normal quiet expiration is largely passive due to lung recoil, when expiration must
       be more forceful (or the lungs are diseased), muscles that increase the abdominal pressure or
       depress the rib cage are enlisted. Provide two examples of muscles that cause abdominal

       pressure to rise. _____

    3. Provide two examples of muscles that depress the rib cage. _____

14. Four nonrespiratory movements are described here. Identify each by inserting
    your answers in the spaces provided.

    1. Sudden inspiration, resulting from spasms of the diaphragm. _____

    2. A deep breath is taken, the glottis is closed, and air is forced out of the lungs against the glottis;

       clears the lower respiratory passageways. _____

    3. As just described, but clears the upper respiratory passageways. _____

    4. Increases ventilation of the lungs; may be initiated by a need to increase oxygen levels in the

       blood. _____

15. The following section concerns respiratory volume measurements. Using key choices, select the terms identified in the following descriptions by inserting the appropriate term or letter in the answer blanks.

KEY CHOICES:

A. Dead space volume   **C.** Inspiratory reserve volume (IRV)   **E.** Tidal volume (TV)

B. Expiratory reserve volume (ERV)   **D.** Residual volume (RV)   **F.** Vital capacity (VC)

_____ **1.** Respiratory volume inhaled or exhaled during normal breathing

_____ **2.** Air in respiratory passages that does not contribute to gas exchange

_____ **3.** Total amount of exchangeable air

_____ **4.** Gas volume that allows gas exchange to go on continuously

_____ **5.** Amount of air that can still be exhaled (forcibly) after a normal exhalation

16. Figure 13-5 is a diagram showing respiratory volumes. Complete the figure by making the following additions.

    1. Bracket the volume representing the vital capacity and color the area yellow; label it VC.

    2. Add green stripes to the area representing the inspiratory reserve volume and label it IRV.

    3. Add red stripes to the area representing the expiratory reserve volume and label it ERV.

    4. Identify and label the respiratory volume, which is *now* yellow. Color the residual volume (RV) blue and label it appropriately on the figure.

    5. Bracket and label the inspiratory capacity (IC).

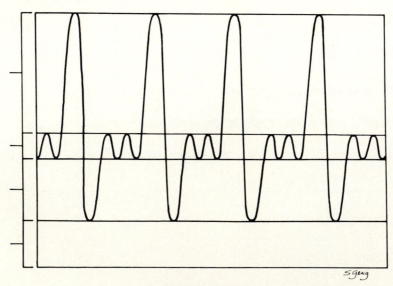

Figure 13-5

17. Use the key choices to correctly complete the following statements, which refer to gas exchanges in the body. Insert the correct letter response in the answer blanks.

    KEY CHOICES:

    A. Active transport   **F.** Diffusion

    B. Air of alveoli to capillary blood   **G.** Higher concentration

    C. Carbon dioxide–poor and oxygen-rich   **H.** Lower concentration

    D. Capillary blood to alveolar air   **I.** Oxygen-poor and carbon dioxide–rich

    E. Capillary blood to tissue cells   **J.** Tissue cells to capillary blood

    _____ 1.

    _____ 2.

    _____ 3.

    _____ 4.

    _____ 5.

    _____ 6.   _____ 8.

    _____ 7.   _____ 9.

    All gas exchanges are made by __(1)__ . When substances pass in this manner, they move from areas of their __(2)__ to areas of their __(3)__ . Thus oxygen continually passes from the __(4)__ and then from the __(5)__ . Conversely, carbon dioxide moves from the __(6)__ and from __(7)__ . From there it passes out of the body during expiration. As a result of such exchanges, arterial blood tends to be __(8)__ while venous blood is __(9)__ .

18. Complete the following statements by inserting your answers in the answer blanks.

    _____ 1.

    _____ 2.

    _____ 3.

    _____ 4.

    Most oxygen is transported bound to __(1)__ inside the red blood cells. Conversely, *most* carbon dioxide is carried in the form of __(2)__ in the __(3)__ . Carbon monoxide poisoning is lethal because carbon monoxide competes with __(4)__ for binding sites.

19. Circle the term that does not belong in each of the following groupings.

    1. ↓ Respiratory rate   ↓ In blood CO_2 Alkalosis   Acidosis

    2. Acidosis   ↑ Carbonic acid   ↓ pH   ↑ pH

    3. Acidosis   Hyperventilation   Hypoventilation CO_2 buildup

    4. Apnea   Cyanosis   ↑ Oxygen   ↓ Oxygen

    5. ↑ Respiratory rate   ↑ Exercise   Anger   ↑ CO_2 in blood

    6. High altitude   ↓ P_{O_2} ↑ P_{CO_2} ↓ Atmospheric pressure

20. There are several types of breathing controls. Match the structures given in Column B to the appropriate descriptions provided in Column A. Place the correct term or letter response in the answer blanks provided.

Column A

_____ , _____   **1.** Respiratory control centers in the pons

_____ , _____   **2.** Respiratory control centers in the medulla

_____   **3.** Respond to overinflation and underinflation of the lungs

_____   **4.** Respond to decreased oxygen levels in the blood

Column B

A. Apneustic center

B. Chemoreceptors in the aortic arch and carotid body

C. Expiratory center

D. Inspiratory center

E. Pneumotaxic center

F. Stretch receptors in the lungs

21. Match the terms in Column B with the pathologic conditions described in Column A

Column A

_____   **1.** Lack or cessation of breathing

_____   **2.** Normal breathing in terms of rate and depth

_____   **3.** Labored breathing, or "air hunger"

_____   **4.** Chronic oxygen deficiency

_____   **5.** Condition characterized by fibrosis of the lungs and an increase in size of the alveolar chambers

_____   **6.** Condition characterized by increased mucus production, which clogs respiratory passageways and promotes coughing

_____   **7.** Respiratory passageways narrowed by bronchiolar spasms

_____   **8.** Together called COPD

_____   **9.** Incidence strongly associated with cigarette smoking; has increased dramatically in women recently

Column B

A. Apnea

B. Asthma

C. Chronic bronchitis

D. Dyspnea

E. Emphysema

F. Eupnea

G. Hypoxia

H. Lung cancer

Developmental Aspects of the Respiratory System

22. Mrs. Jones gave birth prematurely to her first child. At birth, the baby weighed
    2 lb 8 oz. Within a few hours, the baby had developed severe dyspnea and was
    becoming cyanotic. Therapy with a positive pressure ventilator was prescribed.
    Answer the following questions related to the situation just described. Place
    your responses in the answer blanks.

    1. The infant's condition is referred to as _____

    _____ .

    2. It occurs because of a relative lack of _____ .

    3. The function of the deficient substance is to _____

    _____ .

    4. Explain what the positive pressure apparatus accomplishes. _____

    _____ .

23. Complete the following statements by inserting your answers in the answer
    blanks.

    _____ **1.**   The respiratory rate of a newborn baby is approximately
                                __(1)__ respirations per minute. In a healthy adult, the
    _____ **2.**   respiratory range is __(2)__ respirations per minute. Most
                                problems that interfere with the operation of the respiratory
    _____ **3.**   system fall into one of the following categories: infections
                                such as pneumonia, obstructive conditions such as __(3)__
    _____ **4.**   and __(4)__ , and/or conditions that destroy lung tissue,
                                such as __(5)__ . With age, the lungs lose their __(6)__ , and the
    _____ **5.**   __(7)__ of the lungs decreases. Protective mechanisms also
                                become less efficient, causing elderly individuals to be more
    _____ **6.**   susceptible to __(8)__ .

    _____ **7.**

    _____ **8.**

The Incredible Journey:
A Visualization Exercise
for the Respiratory System

You carefully begin to pick your way down, using cartilages as steps.

24. Where necessary, complete statements by inserting the missing word(s) in the answer blanks.

_____ 1.

_____ 2.

_____ 3.

_____ 4.

_____ 5.

_____ 6.

_____ 7.

_____ 8.

_____ 9.

_____ 10.

_____ 11.

_____ 12.

_____ 13.

_____ 14.

_____ 15.

_____ 16.

_____ 17.

_____ 18.

_____ 19.

_____ 20.

Your journey through the respiratory system is to be on foot. To begin, you simply will walk into your host's external nares. You are miniaturized, and your host is sedated lightly to prevent sneezing during your initial observations in the nasal cavity and subsequent descent.

You begin your exploration of the nasal cavity in the right nostril. One of the first things you notice is that the chamber is very warm and humid. High above, you see three large, round lobes, the __(1)__, which provide a large mucosal surface area for warming and moistening the entering air. As you walk toward the rear of this chamber, you see a large lumpy mass of lymphatic tissue, the __(2)__ in the __(3)__, or first portion of the pharynx. As you peer down the pharynx, you realize that it will be next to impossible to maintain your footing during the next part of your journey. It is nearly straight down, and the __(4)__ secretions are like grease. You sit down and dig your heels in to get started. After a quick slide, you land abruptly on one of a pair of flat, sheetlike structures that begin to vibrate rapidly, bouncing you up and down helplessly. You are also conscious of a rhythmic hum during this jostling, and you realize that you have landed on a __(5)__. You pick yourself up and look over the superior edge of the __(6)__, down into the seemingly endless esophagus behind. You chastize yourself for not remembering that the __(7)__ and respiratory pathways separate at this point. Hanging directly over your head is the leaflike __(8)__ cartilage. Normally, you would not have been able to get this far because it would have closed off this portion of the respiratory tract. With your host sedated, however, that protective reflex did not work.

You carefully begin to pick your way down, using the cartilages as steps. When you reach the next respiratory organ, the __(9)__, your descent becomes much easier, because the structure's C-shaped cartilages form a ladderlike supporting structure. As you climb down the cartilages, your face is stroked rhythmically by soft cellular extensions, or __(10)__. You remember that their function is to move mucus laden with bacteria or dust and other debris toward the __(11)__.

You finally reach a point where the descending passageway splits into two __(12)__ , and since you want to control your progress (rather than slide downward), you choose the more horizontal __(13)__ branch. If you remain in the superior portion of the lungs, your return trip will be less difficult because the passageways will be more horizontal than steeply vertical. The passageways get smaller and smaller, slowing your progress. As you are squeezing into one of the smallest of the respiratory passageways, a __(14)__ , you see a bright spherical chamber ahead. You scramble into this __(15)__ , pick yourself up, and survey the area. Scattered here and there are lumps of a substance that looks suspiciously like coal, reminding you that your host is a smoker. As you stand there, a soft rustling wind seems to flow in and out of the chamber. You press your face against the transparent chamber wall and see disklike cells, __(16)__ , passing by in the capillaries on the other side. As you watch, they change from a somewhat bluish color to a bright __(17)__ color as they pick up __(18)__ and unload __(19)__ .

You record your observations and then contact headquarters to let them know you are ready to begin your ascent. You begin your return trek, slipping and sliding as you travel. By the time you reach the inferior edge of the trachea, you are ready for a short break. As you rest on the mucosa, you begin to notice that the air is becoming close and very heavy. You pick yourself up quickly and begin to scramble up the trachea. Suddenly and without warning, you are hit by a huge wad of mucus and catapulted upward and out on to your host's freshly pressed handkerchief! Your host has assisted your exit with a __(20)__ .

Notes

Digestive System and Body Metabolism

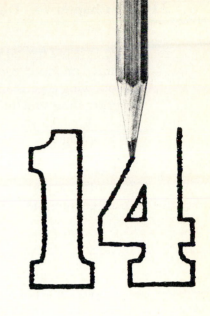

The digestive system processes food so that it can be absorbed and utilized by the body's cells. The digestive organs are responsible for food ingestion, digestion, absorption, and elimination of undigested remains from the body. In one sense, the digestive tract can be viewed as a disassembly line in which food is carried from one stage of its breakdown process to the next by muscular activity, and its nutrients are made available en route to the cells of the body. In addition, the digestive system provides for one of life's greatest pleasures—eating.

The anatomy of both alimentary canal and accessory digestive organs, mechanical and enzymatic breakdown, and absorption mechanisms are covered in this chapter. Important understandings of cellular metabolism (utilization of foodstuffs by body cells) are also considered in this chapter review.

Anatomy of the Digestive System

1. Complete the following statements by inserting your answers in the answer blanks.

_____ 1.

_____ 2.

_____ 3.

_____ 4.

_____ 5.

_____ 6.

_____ 7.

The digestive system is responsible for many body processes. Its functions begin when food is taken into the mouth, or **(1)** . The process called **(2)** occurs as food is broken down both chemically and mechanically. For the broken-down foods to be made available to the body cells, they must be absorbed through the digestive system walls into the **(3)** . Undigestible food remains are removed, or **(4)** , from the body via the **(5)** . The organs forming a continuous tube from the mouth to the anus are collectively called the **(6)** . Organs located outside the digestive proper, which secrete their products into the digestive tract, are referred to as **(7)** digestive system organs.

2.   Figure 14-1 is a frontal view of the digestive system. First, correctly identify all
      structures provided with leader lines. Then select a different color for each of
      the following organs and use it to color the coding circles and the
      corresponding structures of the figure.

      ◯ Esophagus   ◯ Pancreas   ◯ Small intestine

      ◯ Liver   ◯ Salivary glands   ◯ Tongue

      ◯ Large intestine

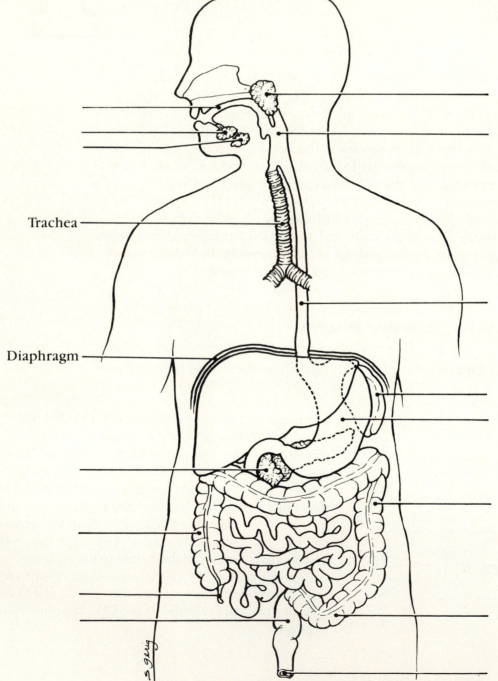

Trachea

Diaphragm

Figure 14-1

3. Figure 14-2 illustrates oral cavity structures. First, correctly identify all structures provided with leader lines. Then color the structure that attaches the tongue to the floor of the mouth red; color the portions of the roof of the mouth unsupported by bone blue; color the structures that are essentially masses of lymphatic tissue yellow; and color the structure that contains the bulk of the taste buds pink.

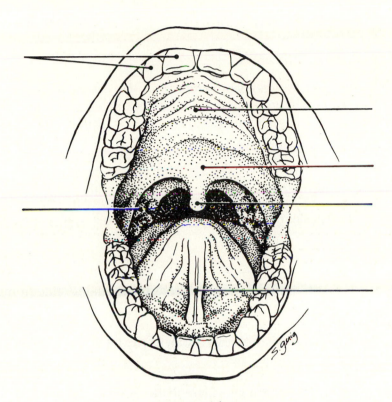

Figure 14-2

4. Various types of glands secrete substances into the alimentary tube. Match the glands listed in Column B to the functions/locations described in Column A. Place the correct term or letter response in the answer blanks.

Column A

_____ 1. Produce mucus; found in the submucosa of the small intestine

_____ 2. Secretion includes amylase, which begins starch digestion in the mouth

_____ 3. Ducts a variety of enzymes in an alkaline fluid into the duodenum

_____ 4. Produces bile, which is transported to the duodenum via the common bile duct

_____ 5. Produces hydrochloric acid and pepsinogen

Column B

A. Brunner's glands

B. Gastric glands

C. Liver

D. Pancreas

E. Salivary glands

5.   Using the key choices, select the terms identified in the following descriptions
      by inserting the appropriate term or letter in the answer blanks.

KEY CHOICES:

A. Anus	**J.** Mesentery	**R.** Rugae
B. Appendix	**K.** Microvilli	**S.** Small intestine
C. Colon	**L.** Oral cavity	**T.** Soft palate
D. Esophagus	**M.** Parietal peritoneum	**U.** Stomach
E. Greater omentum	**N.** Peyer's patches	**V.** Tongue
F. Hard palate	**O.** Pharynx	**W.** Vestibule
G. Haustra	**P.** Pilcae circulares	**X.** Villi
H. Ileocecal valve	**Q.** Pyloric valve	**Y.** Visceral peritoneum
I. Lesser omentum		

_____   **1.** Structure that suspends the small intestine from the posterior body
                                    wall

_____   **2.** Fingerlike extensions of the intestinal mucosa that increase the
                                    surface area

_____   **3.** Collections of lymphatic tissue found in the submucosa of the
                                    small intestine

_____   **4.** Folds of the small intestine wall

_____ , _____   **5.** Two anatomic regions involved in the mechanical breakdown of
                                    food

_____   **6.** Organ that mixes food in the mouth and initiates swallowing

_____   **7.** Common passage for food and air

_____ , _____ , _____   **8.** Three extensions/modifications of the peritoneum

_____   **9.** Literally a food chute; has no digestive or absorptive role

_____   **10.** Folds of the stomach mucosa

_____   **11.** Saclike outpocketings of the large intestine wall

_____   **12.** Projections of the plasma membrane of a cell that increase the
                                    cell's surface area

_____ 13. Prevents food from moving back into the small intestine once it has entered the large intestine

_____ 14. Organ responsible for most food and water absorption

_____ 15. Organ primarily involved in water absorption and feces formation

_____ 16. Area between the teeth and lips/cheeks

_____ 17. Blind sac hanging from the cecum

_____ 18. Organ in which protein digestion begins

_____ 19. Membrane attached to the lesser curvature of the stomach

_____ 20. Organ into which the stomach empties

_____ 21. Sphincter, controlling the movement of food from the stomach into the duodenum

_____ 22. Uvula hangs from its posterior edge

_____ 23. Organ that receives pancreatic juice and bile

_____ 24. Serosa of the abdominal cavity wall

_____ 25. Major site of vitamin (K, B) formation by bacteria

_____ 26. Region, containing two sphincters, through which feces are expelled from the body

_____ 27. Anteriosuperior boundary of the oral cavity; supported by bone

_____ 28. Extends as a double fold from the greater curvature of the stomach

_____ 29. Superiorly its muscle is striated; inferiorly it is smooth

6. Figure 14-3, A, is a longitudinal section of the stomach. First, use the following terms to identify the regions provided with leader lines on the figure.

Body   Pyloric region   Greater curvature   Cardioesophageal valve

Fundus   Pyloric valve   Lesser curvature

Then select different colors for each of the following structures/areas and use them to color the coding circles and corresponding structures/areas on the figure.

◯ Oblique muscle layer   ◯ Longitudinal muscle layer   ◯ Circular muscle layer

◯ Area where rugae are visible   ◯ Serosa

Figure 14-3, B, shows two types of secretory cells found in gastric glands. Color the hydrochloric acid–secreting cells red and color the cells that produce protein-digesting enzymes blue.

A

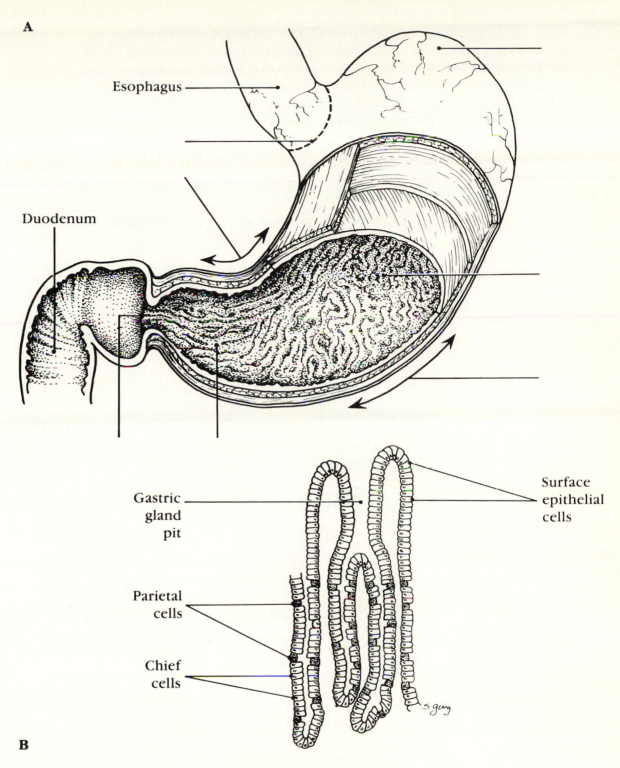

Esophagus

Duodenum

Gastric
gland
pit

Surface
epithelial
cells

Parietal
cells

Chief
cells

B

Figure 14-3

7. The walls of the alimentary canal have four typical layers, as illustrated in Figure 14-4. Identify each layer by placing its correct name in the space before the appropriate description. Then select different colors for each layer and use them to color the coding circles and corresponding structures on the figure. Finally, assume the figure shows a cross-sectional view of the small intestine and label the three structures provided with leader lines.

_____ ◯ **1.** The secretory and absorptive layer

_____ ◯ **2.** Layer composed of at least two muscle layers

_____ ◯ **3.** Connective tissue layer, containing blood, lymph vessels, and nerves

_____ ◯ **4.** Outermost layer of the wall; visceral peritoneum

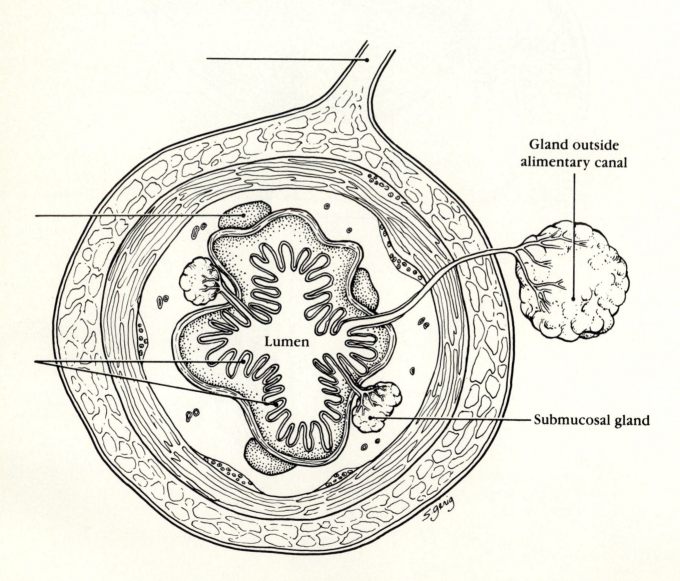

Figure 14-4

8. Figure 14-5 shows three views of the small intestine. First, label the villi in views B and C and the plicae circulares in views A and B. Then select different colors for each term listed below and use them to color in the coding circles and corresponding structures in view C.

◯ Surface epithelium   ◯ Lacteal   ◯ Capillary network

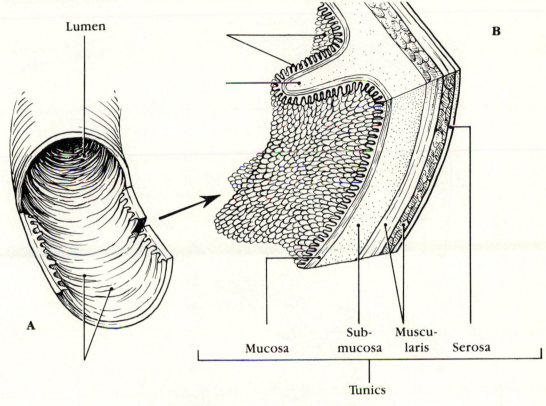

Lumen

A

B

Mucosa   Sub-mucosa   Muscu-laris   Serosa

Tunics

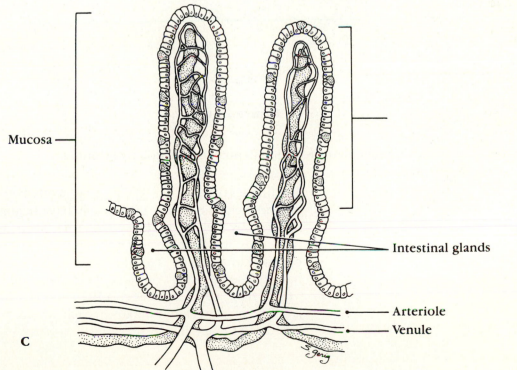

Mucosa

Intestinal glands

Arteriole

Venule

C

Figure 14-5

9. Three accessory organs are illustrated in Figure 14-6. Identify each of the three organs and the ligament provided with leader lines on the figure. Then select different colors for the following structures and use them to color the coding circles and the corresponding structures on the figure.

○ Common hepatic duct   ○ Common bile duct

○ Cystic duct   ○ Pancreatic duct

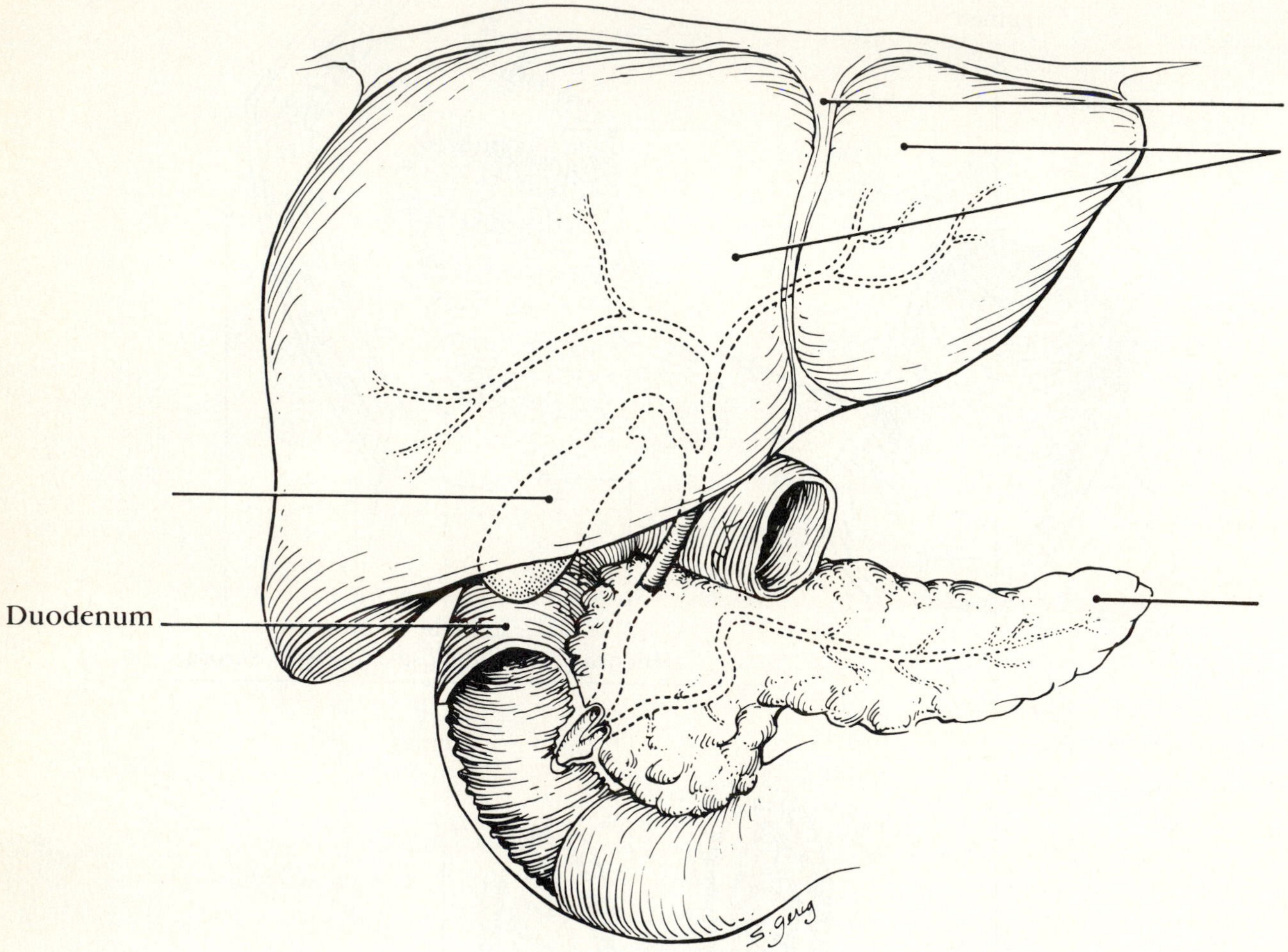

Duodenum _____

Figure 14-6

10. Complete the following statements referring to human dentition by inserting your answers in the answer blanks.

_____ 1.

_____ 2.

_____ 3.

_____ 4.

The first set of teeth, called the __(1)__ teeth, begin to appear around the age of __(2)__ and usually have begun to be replaced by the age of __(3)__. The __(4)__ teeth are more numerous; that is, there are __(5)__ teeth in the second set as opposed to a total of __(6)__ teeth in the first set. If an adult has a full set of teeth, you can expect to find two __(7)__, one __(8)__, two __(9)__, and three __(10)__ in one side of each jaw. The most posterior molars in each jaw are commonly called __(11)__ teeth.

_____ **5.**   _____ **9.**

_____ **6.**   _____ **10.**

_____ **7.**   _____ **11.**

_____ **8.**

11. First, use the key choices to identify each tooth area described below and to label the tooth diagrammed in Figure 14-7. Second, select different colors to represent the key choices and use them to color in the coding circles and corresponding structures in the figure. Third, add labels to the figure to identify the crown, gingiva, and root of the tooth.

KEY CHOICES:

A. Cementum   **C.** Enamel   **E.** Pulp

B. Dentin   **D.** Periodontal membrane

_____ **1.** Material covering the tooth root

_____ **2.** Hardest substance in the body; covers tooth crown

_____ **3.** Attaches the tooth to bone and surrounding alveolar structures

_____ **4.** Forms the bulk of tooth structure; similar to bone

_____ **5.** A collection of blood vessels, nerves, and lymphatics

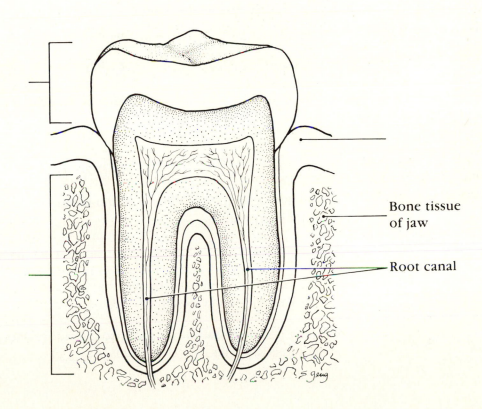

Bone tissue of jaw

Root canal

Figure 14-7

12. Circle the term that does not belong in each of the following groupings.

   1. Nasopharynx   Esophagus   Laryngopharynx   Oropharynx

   2. Villi   Plicae circulares   Rugae   Microvilli

   3. Salivary glands   Pancreas   Liver   Gallbladder

   4. Duodenum   Cecum   Jejunum   Ileum

   5. Ascending colon   Haustra   Circular folds   Cecum

   6. Mesentery   Frenulum   Greater omentum   Parietal peritoneum

   7. Parotid   Sublingual   Submandibular   Palatine

   8. Protein-digesting enzymes   Saliva   Intrinsic factor   HCl

   9. Colon   Water absorption   Protein absorption   Vitamin B absorption

Digestive System Function: Food Movement, Breakdown, and Absorption

13. This section relates to food breakdown in the digestive tract. Using key choices, select the appropriate terms to complete the following statements. Insert the correct letter or term in the answer blanks.

   KEY CHOICES:

A. Bicarbonate-rich fluid	**F.** HCl	**K.** Mucus
B. Bile	**G.** Hormonal stimulus	**L.** Pepsin
C. Chewing	**H.** Lipases	**M.** Psychologic stimulus
D. Churning	**I.** Mechanical stimulus	**N.** Rennin
E. Disaccharases	**J.** Mouth	**O.** Salivary amylase

   _____ **1.** Starch digestion begins in the mouth when __(1)__ is ducted in by the salivary glands.

   _____ **2.** The means of mechanical food breakdown in the mouth is __(2)__ .

   _____ **3.** The fact that the mere thought of a relished food can make your mouth water is an example of __(3)__ .

   _____ **4.** Many people chew gum to increase saliva formation when their mouth is dry. This type of stimulus is a __(4)__ .

_____   **5.** Protein foods are largely acted on in the stomach by __(5)__ .

_____   **6.** For the stomach protein-digesting enzymes to become active, __(6)__ is needed.

_____   **7.** Since living cells of the stomach (and everywhere) are largely protein, it is amazing that they are not digested by the activity of stomach enzymes. The most important means of stomach protection is the __(7)__ it produces.

_____   **8.** A milk protein-digesting enzyme found in children but uncommon in adults is __(8)__ .

_____   **9.** The third layer of smooth muscle found in the stomach wall allows mixing and mechanical breakdown by __(9)__ .

_____   **10.** Important intestinal enzymes are the __(10)__ .

_____   **11.** The small intestine is protected from the corrosive action of hydrochloric acid in chyme by __(11)__ , which is ducted in by the pancreas.

_____   **12.** The pancreas produces protein-digesting enzymes, amylase, and nucleases. It is the only important source of __(12)__ .

_____   **13.** A nonenzyme substance that causes fat to be dispersed into smaller globules is __(13)__ .

14. Identify the pathologic conditions described below by using terms from the key choices. Insert the correct term or letter in the answer blanks.

KEY CHOICES:

A. Appendicitis	**C.** Diarrhea	**E.** Heartburn	**G.** Peritonitis
B. Constipation	**D.** Gallstones	**F.** Jaundice	**H.** Ulcer

_____   **1.** Inflammation of the abdominal serosa

_____   **2.** Condition resulting from the reflux of acidic gastric juice into the esophagus

_____   **3.** Usually indicates liver problems or blockage of the biliary ducts

_____   **4.** An erosion of the stomach or duodenal mucosa

_____   **5.** Passage of watery stools

_____   **6.** Causes severe epigastric pain; associated with prolonged storage of bile in the gallbladder

_____   **7.** Inability to pass feces; often a result of poor bowel habits

15. Hormonal stimuli are important in digestive activities that occur in the stomach and small intestine. Using the key choices, identify the hormones that function as described in the following statements. Insert the correct term or letter response in the answer blanks.

KEY CHOICES:

A. Cholecystokinin   **B.** Gastrin   **C.** Secretin

_____ , _____   **1.** These two hormones stimulate the pancreas to release its secretions.

_____   **2.** This hormone stimulates increased production of gastric juice.

_____   **3.** This hormone causes the gallbladder to release stored bile.

_____   **4.** This hormone causes the liver to increase its output of bile.

16. Various types of foods are ingested in the diet and broken down to their building blocks. Use the key choices to complete the following statements according to these understandings. Insert the correct letter(s) in the answer blanks. In most cases, more than one choice applies.

KEY CHOICES:

A. Amino acids   **E.** Fatty acids   **H.** Glucose   **K.** Meat/fish

B. Bread/pasta   **F.** Fructose   **I.** Lactose   **L.** Starch

C. Cheese/cream   **G.** Galactose   **J.** Maltose   **M.** Sucrose

D. Cellulose

_____ , _____   **1.** Examples of carbohydrate foods in the diet.

_____ , _____ , _____   **2.** The building blocks of carbohydrates are monosaccharides, or simple sugars. The three common simple sugars in our diet are ___, ___, and ___.

_____   **3.** Of the simple sugars, ___ is most important because it is the sugar referred to as "blood sugar."

_____ , _____ , _____   **4.** Disaccharides include ___, ___, and ___.

_____   **5.** The only important *digestible* polysaccharide is ___.

_____   **6.** An indigestible polysaccharide that aids elimination because it adds bulk to the diet is ___.

_____ , _____   **7.** Protein-rich foods include ___ and ___.

_____ **8.** Protein foods must be digested to ___ before they can be absorbed.

_____ **9.** Fatty foods ingested in the normal diet include ___.

_____ **10.** Fats are broken down to two types of building blocks, ___ and glycerol.

17. Dietary substances capable of being absorbed are listed next. If the substance is _most often_ absorbed from the digestive tract by active transport processes, put an _A_ in the blank. If it is usually absorbed passively (by diffusion or osmosis), put a _P_ in the blank. In addition, circle the substance that is _most likely_ to be absorbed into a lacteal rather than into the capillary bed of the villus.

_____ **1.** Water   _____ **3.** Simple sugars   _____ **5.** Electrolytes

_____ **2.** Amino acids   _____ **4.** Fatty acids

18. Complete the following statements that describe mechanisms of food mixing and movement. Insert your responses in the answer blanks.

_____ **1.**

_____ **2.**

_____ **3.**

_____ **4.**

_____ **5.**

_____ **6.**

_____ **7.**

_____ **8.**

_____ **9.**

_____ **10.**

_____ **11.**

_____ **12.**

_____ **13.**

_____ **14.**

_____ **15.**

_____ **16.**

_____ **17.**

Swallowing, or __(1)__, occurs in two major phases—the __(2)__ and __(3)__. During the voluntary phase, the __(4)__ is used to push the food into the throat, and the __(5)__ rises to close off the nasal passageways. As food is moved involuntarily through the pharynx, the __(6)__ rises to ensure that its passageway is covered by the __(7)__, so that ingested substances do not enter respiratory passages. It is possible to swallow water while standing on your head because the water is carried along the esophagus involuntarily by the process of __(8)__. The pressure exerted by food on the __(9)__ valve causes it to open so that food can enter the stomach.

The two major types of movements that occur in the small intestine are __(10)__ and __(11)__. One of these movements, the __(12)__, acts to continually mix the food with digestive juices, and (strangely) also plays a major role in propelling foods along the tract. Still another type of movement seen only in the large intestine, __(13)__, occurs infrequently and acts to move feces over relatively long distances toward the anus. Presence of feces in the __(14)__ excites stretch receptors so that the __(15)__ reflex is initiated. Irritation of the gastrointestinal tract by drugs or bacteria might stimulate the __(16)__ center in the medulla, causing __(17)__, which is essentially a reverse peristalsis.

Metabolism

19. Using the key choices, identify the foodstuffs used by cells in the cellular
functions described below. Insert the correct term or key letter in the answer
blanks.

KEY CHOICES:

A. Amino acids   **B.** Carbohydrates   **C.** Fats

_____   **1.** The most used substance for producing the energy-rich ATP

_____   **2.** Important in building myelin sheaths and cell membranes

_____   **3.** Tend to be conserved by cells

_____   **4.** The second most important food source for making cellular energy.

_____   **5.** Form insulating deposits around body organs and beneath the skin

_____   **6.** Used to make the bulk of cell structure and functional
substances such as enzymes

20. The liver has many functions. In addition to its digestive function, it is the major
metabolic organ of the body. Complete the following statements that expand
on the liver's function by inserting the correct terms in the answer blanks.

_____ **1.**   Metabolism is the total of all __(1)__ reactions that occur in
the body. Generally, reactions in which substances are
_____ **2.**   broken down into simpler substances are called __(2)__
reactions, whereas constructive reactions are called __(3)__ .
_____ **3.**   In its metabolic role, the liver uses amino acids from the
nutrient-rich hepatic portal blood to make many blood
_____ **4.**   proteins such as __(4)__ , which helps to hold water in the
bloodstream, and __(5)__ , which prevent blood loss when
_____ **5.**   blood vessels are damaged. The liver also makes a steroid
substance that is released to the blood. This steroid, __(6)__ ,
_____ **6.**   has been implicated in high blood pressure and heart
disease. The liver also acts to maintain constant blood
_____ **7.**   glucose levels. It removes glucose from the blood when
blood glucose levels are high, a condition called __(7)__ ,
_____ **8.**   and stores it as __(8)__ . Then, when blood glucose levels are
low, a condition called __(9)__ , liver cells break down the
_____ **9.**   stored carbohydrate and release glucose to the blood once
again. This latter process is termed __(10)__ . When the liver
_____ **10.**   makes glucose from noncarbohydrate substances such as
fats or proteins, the process is termed __(11)__ . In addition
_____ **11.**   to its processing of amino acids and sugars, the liver plays

_____ 10.

_____ 11.

an important role in the processing of fats. Other functions of the liver include the __(12)__ of drugs and alcohol, and its __(13)__ cells protect the body by ingesting bacteria and other debris.

21. Circle the term that does not belong in each of the following groupings.

1. BMR   TMR   Rest   Postabsorptive state

2. Thyroxine   Iodine   ↓ Metabolic rate   ↑ Metabolic rate

3. Obese person   ↓ Metabolic rate   Women   Child

4. 4 Kcal/gram   Fats   Carbohydrates   Proteins

22. Using the key choices, select the terms identified in the following descriptions. Insert the appropriate term or letter in the answer blanks.

KEY CHOICES:

A. Blood

B. Constriction of skin blood vessels

C. Frostbite

D. Heat

E. Hyperthermia

F. Hypothalamus

G. Hypothermia

H. Perspiration

I. Radiation

J. Pyrogens

K. Shivering

_____ **1.** Byproduct of cell metabolism

_____ , _____ **2.** Means of conserving/increasing body heat

_____ **3.** Means by which heat is distributed to all body tissues

_____ **4.** Site of the body's thermostat

_____ **5.** Chemicals released by injured tissue cells and bacteria, causing resetting of the thermostat

_____ **6.** Death of cells deprived of oxygen and nutrients, resulting from withdrawal of blood from the skin circulation

_____ , _____ **7.** Means of liberating excess body heat

_____ **8.** Extremely low body temperature

_____ **9.** Fever

23. This section considers the process of cellular respiration. Insert the correct *word(s)* from the key choices in the answer blanks.

KEY CHOICES:

A. ATP	**G.** Basal metabolic rate (BMR)	**M.** Ketosis
B. Acetic acid	**H.** Carbon dioxide	**N.** Monosaccharides
C. Acetoacetic acid	**I.** Essential	**O.** Oxygen
D. Acetone	**J.** Fatty acids	**P.** Total metabolic rate (TMR)
E. Amino acids	**K.** Glucose	**Q.** Urea
F. Ammonia	**L.** Glycogen	**R.** Water

_____ 1.

_____ 2.

_____ 3.

_____ 4.

_____ 5.

_____ 6.

_____ 7.

_____ 8.

_____ 9.

_____ 10.

_____ 11.

_____ 12.

The key "fuel" used by body cells is __(1)__ . The cells break this fuel molecule apart piece by piece. The hydrogen removed is combined with __(2)__ to form __(3)__ , while its carbon leaves the body in the form of __(4)__ gas. The importance of this process is that it provides __(5)__ , a form of energy that the cells can use to power all their activities. For carbohydrates to be oxidized, or burned for energy, they must first be broken down to __(6)__ . When carbohydrates are unavailable to prime the metabolic pump, intermediate products of fat metabolism such as __(7)__ and __(8)__ accumulate in the blood, causing __(9)__ and low blood pH. Amino acids are actively accumulated by cells because protein cannot be made unless all amino acid types are present. The amino acids that *must* be taken in the diet are called __(10)__ amino acids. When amino acids are oxidized to form cellular energy, their amino groups are removed and liberated as __(11)__ . In the liver, this is combined with carbon dioxide to form __(12)__ , which is removed from the body by the kidneys.

Developmental Aspects of the Digestive System

24. Using the key choices, select the terms identified in the following descriptions.
    Insert the correct term or letter in the answer blanks.

    KEY CHOICES:

A. Accessory organs	**F.** Gallbladder problems	**K.** Rooting
B. Alimentary canal	**G.** Gastritis	**L.** Sucking
C. Appendicitis	**H.** PKU	**M.** Stomach
D. Cleft palate/lip	**I.** Periodontal disease	**N.** Tracheoesophageal fistula
E. Cystic fibrosis	**J.** Peristalsis	**O.** Ulcers

    _____ **1.** Internal tubelike cavity of the embryo

    _____ **2.** Glands that branch out from the digestive mucosa

    _____ **3.** Most common congenital defect; aspiration of feeding common

    _____ **4.** Congenital condition characterized by a connection between digestive and respiratory passageways

    _____ **5.** Congenital condition in which large amounts of mucus are produced, clogging respiratory passageways and pancreatic ducts

    _____ **6.** Metabolic disorder characterized by an inability to properly use the amino acid phenylalanine

    _____ **7.** Reflex aiding the newborn baby to find the nipple

    _____ **8.** Vomiting is common in infants because this structure is small

    _____ **9.** Most common adolescent digestive system problem

    _____ , _____ **10.** Inflammations of the gastrointestinal tract

    _____ **11.** Condition of loose teeth and inflamed gums; generally seen in elderly people

    _____ **12.** In old age, the production of digestive juices decreases and this action slows, leading to elimination problems

The Incredible Journey:
A Visualization Exercise
for the Digestive System

*. . . the passage beneath you opens, and you fall into
a huge chamber with mountainous folds.*

25.   Where necessary, complete statements by inserting the missing word(s) in the
answer blanks.

_____ 1.

_____ 2.

_____ 3.

_____ 4.

_____ 5.

_____ 6.

_____ 7.

_____ 8.

_____ 9.

_____ 10.

_____ 11.

_____ 12.

_____ 13.

_____ 14.

_____ 15.

_____ 16.

_____ 17.

In this journey you are to travel through the digestive tract
as far as the appendix and then await further instructions.
You are miniaturized as usual and provided with a wet suit
to protect you from being digested during your travels.
You have a very easy entry into your host's open mouth.
You look around and notice the glistening pink lining,
or __(1)__ , and the perfectly cared-for teeth. Within a few
seconds, the lips part and you find yourself surrounded by
bread. You quickly retreat to the safety of the __(2)__
between the teeth and the cheek to prevent getting
chewed. From there you watch with fascination as a
number of openings squirt fluid into the chamber, and the
__(3)__ heaves and rolls, mixing the bread with the fluid. As
the bread begins to disappear, you decide that the fluid
contains the enzyme __(4)__ . You then walk toward the
back of the oral cavity. Suddenly, you find yourself being
carried along by a squeezing motion of the walls around
you. The name given to this propelling motion is __(5)__ .
As you are carried helplessly downward, you see two
openings—the __(6)__ and the __(7)__ —below you. Just as
you are about to straddle the solid area between them to
stop your descent, the structure to your left moves quickly
upward, and a trapdoor-like organ, the __(8)__ , flaps over
its opening. Down you go in the dark, seeing nothing.
Then the passage beneath you opens, and you fall into a
huge chamber with mountainous folds. Obviously, you
have reached the __(9)__ . The folds are very slippery, and
you conclude that it must be the __(10)__ coat that you read
about earlier. As you survey your surroundings, juices
begin to gurgle into the chamber from pits in the "floor,"
and your face begins to sting and smart. You cannot seem
to escape this caustic fluid and conclude that it must be
very dangerous to your skin since it contains __(11)__
and __(12)__ . You reach down and scoop up some of the
slippery substance from the folds and smear it on your

face, confident that if it can protect this organ it can protect you as well! Relieved, you begin to slide toward the organ's far exit and squeeze through the tight __(13)__ valve into the next organ. In the dim light, you see lumps of cellulose lying at your feet and large fat globules dancing lightly about. A few seconds later, your observations are interrupted by a wave of fluid, pouring into the chamber from an opening high in the wall above you. The large fat globules begin to fall apart, and you decide that this enzyme flood has to contain __(14)__, and that the opening must be the duct from the __(15)__. As you move quickly away to escape the deluge, you lose your footing and find yourself on a rollercoaster ride—twisting, coiling, turning, and diving through the lumen of this active organ. As you move, you are stroked by velvety, fingerlike projections of the wall, the __(16)__. Abruptly your ride comes to a halt as you are catapulted through the __(17)__ valve and fall into the appendix. Headquarters informs you that you are at the end of your journey. Your exit now depends on your own ingenuity.

Urinary System

Metabolism of nutrients by body cells produces various wastes such as carbon dioxide and nitrogenous wastes (creatinine, urea, and ammonia), as well as imbalances of water and essential ions. The metabolic wastes and excesses must be eliminated from the body. Essential substances are retained to ensure internal homeostasis and proper body functioning. Although several organ systems are involved in excretory processes, the urinary system is primarily responsible for the removal of the nitrogenous wastes from the blood. In addition to this purely excretory function, the kidneys maintain the electrolyte, acid–base, and fluid balances of the blood. Thus, kidneys are major homeostatic organs of the body. Malfunction of the urinary system, particularly the kidneys, leads to a failure of homeostasis, resulting (unless corrected) in death.

Student activities in this chapter are concerned with identification of urinary system structures, composition of urine, and the physiological processes involved in urine formation.

1. Complete the following statements by inserting your answers in the answer blanks.

1. _____

2. _____

3. _____

4. _____

5. _____

6. _____

7. _____

8. _____ 9. _____ 10.

The kidney is referred to as an excretory organ because it excretes **(1)** wastes. It is also a major homeostatic organ because it maintains the electrolyte, **(2)** , and **(3)** balance of the blood. Urine is continuously formed by the **(4)** and is routed down the **(5)** by the mechanism of **(6)** to a storage organ called the **(7)** . Eventually the urine is conducted to the body exterior by the **(8)** . In males, this tubelike structure is about **(9)** inches long; in females, it is approximately **(10)** inches long.

Anatomy

2. Figure 15-1 is a front view of the entire urinary system. Identify and select
    different colors for the following organs. Use them to color the coding circles
    and the corresponding organs on the figure.

◯ Kidney   ◯ Bladder   ◯ Ureters   ◯ Urethra

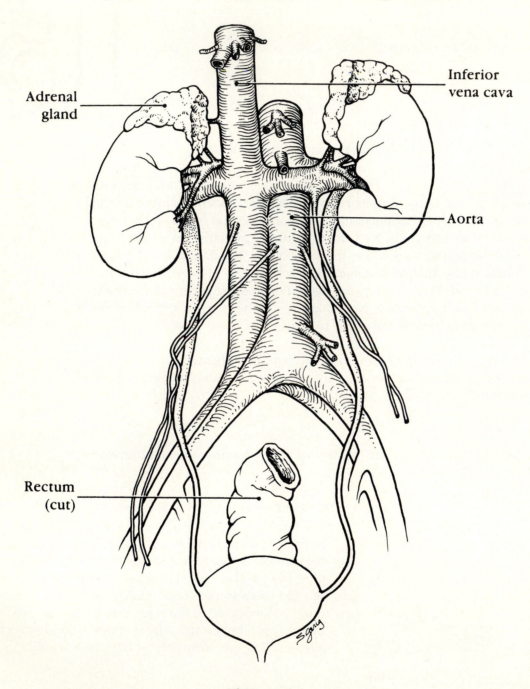

Figure 15-1

3. Figure 15-2 is a longitudinal section of a kidney. First, using the correct anatomic terminology, label the following regions/structures indicated by leader lines on the figure.

    - Fibrous membrane, surrounding the kidney

    - Basinlike area of the kidney that is continuous with the ureter

    - Cuplike extension of the pelvis that drains the apex of a pyramid

    - Area of cortical tissue running through the medulla

    Then, excluding the color red, select different colors to identify the following areas and structures. Then color in the coding circles and the corresponding area/structures on the figure; label these regions using the correct anatomic terms.

    ○ Area of the kidney that contains the greatest proportion of nephron structures

    ○ Striped-appearing structures formed primarily of collecting ducts

    Finally, beginning with the renal artery, *draw in* the vascular supply to the cortex on the figure. Include and label the interlobar artery, arcuate artery, and interlobular artery. Color the vessels bright red.

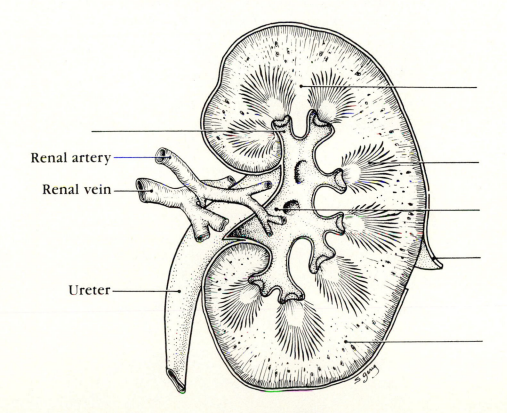

Renal artery

Renal vein

Ureter

Figure 15-2

4.   Circle the term that does not belong in each of the following groupings.

  1. Bladder   Kidney   Transitional epithelium   Detrusor muscle

  2. Trigone   Ureter openings   Urethral opening   Bladder   Forms urine

  3. Surrounded by prostate gland   Contains internal and external sphincters
       Continuous with renal pelvis   Urethra   Drains bladder

  4. Juxtaglomerular apparatus   Distal tubule   Glomerulus   Afferent arteriole

  5. Glomerulus   Peritubular capillaries   Blood vessels   Collecting tubule

  6. Cortical nephrons   Juxtamedullary nephrons   Cortex/medulla junction
       Long loops of Henle

  7. Nephron   Proximal convoluted tubule   Distal convoluted tubule
       Collecting duct

  8. Medullary pyramids   Glomeruli   Renal pyramids   Collecting tubules

5.   Figure 15-3 is a diagram of the nephron and associated blood supply. First,
      match each of the numbered structures on the figure to one of the terms below
      the figure. Place the terms in the numbered spaces provided below. Then color
      the structure on the figure that contains podocytes green; the filtering apparatus
      red; the capillary bed that directly receives the reabsorbed substances from the
      tubule cells blue; the structure into which the nephron empties its urine product
      yellow; and the tubule area that is the primary site of tubular reabsorption
      orange.

  _____ **1.**   _____ **9.**

  _____ **2.**   _____ **10.**

  _____ **3.**   _____ **11.**

  _____ **4.**   _____ **12.**

  _____ **5.**   _____ **13.**

  _____ **6.**   _____ **14.**

  _____ **7.**   _____ **15.**

  _____ **8.**

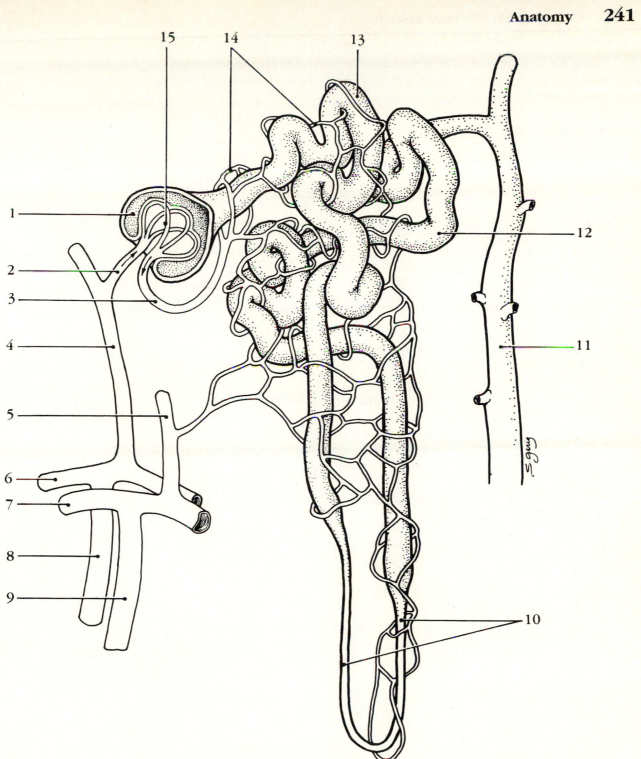

Figure 15-3

Afferent arteriole	DCT	Interlobular artery
Arcuate artery	Efferent arteriole	Interlobular vein
Arcuate vein	Glomerulus	Loop of Henle
Bowman's capsule	Interlobar artery	PCT
Collecting tubule	Interlobar vein	Peritubular capillaries

6. Using the key choices, identify the structures that best fit the following descriptions. Insert the correct term or corresponding letter in the answer blanks.

KEY CHOICES:

A. Bladder   **B.** Urethra   **C.** Ureter

_____ **1.** Drains the bladder

_____ **2.** Storage area for urine

_____ **3.** Contains the trigone

_____ **4.** Surrounded by the prostate gland in males

_____ **5.** Conducts urine by peristalsis

_____ **6.** Substantially longer in males than in females

_____ **7.** Continuous with the renal pelvis

_____ **8.** Contains transitional epithelium

_____ **9.** Contains internal and external sphincters

_____ **10.** Also transports sperm in males

7. Match the terms in Column B with the appropriate descriptions provided in Column A. Insert the correct term or letter in the answer blanks.

Column A

_____ **1.** Inflammatory condition particularly common in women who practice poor toileting habits

_____ **2.** Backup of urine into the kidney; often a result of a blocked ureter

_____ **3.** Toxic condition due to renal failure

_____ **4.** Inflammation of a kidney

_____ **5.** A condition in which excessive amounts of urine are produced because of a deficiency of antidiuretic hormone (ADH)

_____ **6.** Dropping of the kidney to a more inferior position in the abdomen; may result from a rapid weight loss that decreases the fatty cushion surrounding the kidney

Column B

A. Cystitis

B. Diabetes insipidus

C. Hydronephrosis

D. Ptosis

E. Pyelonephritis

F. Uremia

Physiology

8. Complete the following statements by inserting your answers in the answer blanks.

_____ 1.

_____ 2.

_____ 3.

_____ 4.

_____ 5.

_____ 6.

_____ 7.

_____ 8.

_____ 9.

_____ 10.

_____ 11.

_____ 12.

_____ 13.

_____ 14.   _____ 17.

_____ 15.   _____ 18.

_____ 16.   _____ 19.

The glomerulus is a unique high-pressure capillary bed, because the __(1)__ arteriole feeding it is larger in diameter than the __(2)__ arteriole draining the bed. Glomerular filtrate is very similar to __(3)__, but it has fewer proteins. Mechanisms of tubular reabsorption include __(4)__ and __(5)__. As an aid for the reabsorption process, the cells of the proximal convoluted tubule have dense __(6)__ on their luminal surface, which increase the surface area dramatically. Other than reabsorption, an important tubule function is __(7)__, which is important for ridding the body of substances not already in the filtrate. Blood composition depends on __(8)__, __(9)__, and __(10)__. In a day's time, 180 liters of blood plasma are filtered into the kidney tubules, but only about __(11)__ liters of urine are actually produced. __(12)__ is responsible for the normal yellow color of urine. The three major nitrogenous wastes found in the blood, which must be disposed of, are __(13)__, __(14)__, and __(15)__. The kidneys are the final "judges" of how much water is to be lost from the body. When water loss via vaporization from the __(16)__, or __(17)__ from the skin is excessive, urine output __(18)__. If the kidneys become nonfunctional, __(19)__ is used to cleanse the blood of impurities.

9. Circle the term that does not belong in each of the following groupings. (BP = Blood pressure.)

1. Hypothalamus   ADH   Aldosterone   Osmoreceptors

2. Glomerulus   Secretion   Filtration   ↑ BP

3. Aldosterone   ↑ Na$^+$ reabsorption   ↑ K$^+$ reabsorption   ↑ BP

4. ADH   ↓ BP   ↑ Blood volume   ↑ Water reabsorption

5. ↓ Aldosterone   Edema   ↓ Blood volume   ↓ K$^+$ Retention

6. ↓ Urine pH   ↑ H$^+$ in urine   ↑ HCO$_3^-$ in urine   ↑ Ketones

10. Decide whether the following conditions would cause urine to become more acidic or more basic. If more acidic, insert an *A* in the blank; if more basic, insert a *B* in the blank.

_____ **1.** Protein-rich diet   _____ **4.** Diabetes mellitus

_____ **2.** Bacterial infection   _____ **5.** Vegetarian diet

_____ **3.** Starvation

11. Decide whether the following conditions would result in an increase or decrease in urine specific gravity. Insert *I* in the answer blank to indicate an increase and *D* to indicate a decrease.

_____ **1.** Drinking excessive fluids   _____ **4.** Using diuretics

_____ **2.** Chronic renal failure   _____ **5.** Limited fluid intake

_____ **3.** Pyelonephritis   _____ **6.** Fever

12. Assuming *normal* conditions, note whether each of the following substances would be *(A)* in greater concentration in the urine than in the glomerular filtrate, *(B)* in lesser concentration in the urine than in the glomerular filtrate, or *(C)* absent in both urine and glomerular filtrate. Place the correct letter in the answer blanks.

_____ **1.** Water   _____ **5.** Glucose   _____ **9.** Potassium ions

_____ **2.** Urea   _____ **6.** Albumin   _____ **10.** Red blood cells

_____ **3.** Uric acid   _____ **7.** Creatinine   _____ **11.** Sodium ions

_____ **4.** Pus (white blood cells)   _____ **8.** Hydrogen ions   _____ **12.** Amino acids

13. Several specific terms are used to indicate the presence of abnormal urine constituents. Identify each of the following abnormalities by inserting the term that names the condition in the spaces provided. Then for each condition, provide one possible cause of the condition in the remaining spaces.

1. Presence of red blood cells: _____. Cause: _____.

2. Presence of ketones: _____. Cause: _____.

3. Presence of albumin: _____. Cause: _____.

4. Presence of pus: _____. Cause: _____.

5. Presence of bile: _____. Cause: _____.

6. Presence of "sand": _____. Cause: _____.

7. Presence of glucose: _____. Cause: _____.

14. Glucose and albumin are both normally absent from urine, but the reason for their exclusion differs. Respond to the following questions in the spaces provided.

    1. Explain the reason for the absence of glucose in urine. _____

    _____

    2. Explain the reason for the absence of albumin in urine. _____

    _____

15. Complete the following statements by inserting your answers in the answer blanks.

    _____ **1.**

    _____ **2.**

    _____ **3.**

    _____ **4.**

    _____ **5.**

    _____ **6.**

    _____ **7.**

    _____ **8.**

    _____ **9.**

    _____ **10.**

    _____ **11.**

    _____ **12.**

    _____ **13.**

Another term that means voiding or emptying of the bladder is __(1)__ . Voiding has both voluntary and involuntary aspects. As urine accumulates in the bladder, __(2)__ are activated. This results in a reflex that causes the muscular wall of the bladder to __(3)__ , and urine is forced past the __(4)__ sphincter. The __(5)__ sphincter is controlled __(6)__ ; thus an individual can temporarily postpone emptying the bladder until it has accumulated __(7)__ ml of urine. __(8)__ is a condition in which voiding cannot be voluntarily controlled. It is normal in __(9)__ , because nervous control of the voluntary sphincter has not been achieved. Other conditions that might result in an inability to control the sphincter include __(10)__ and __(11)__ . __(12)__ is essentially the opposite of incontinence and often is a problem in elderly men due to __(13)__ enlargement.

16. Renin release by the JG apparatus of the nephrons plays an important role in increasing systemic blood pressure. Check (✓) all factors that promote renin release.

    _____ **1.** Drop in systemic blood pressure

    _____ **2.** Increased stretch of the afferent arterioles

    _____ **3.** Sympathetic nervous system stimulation of the JG cells

    _____ **4.** Low filtrate osmolarity

17. Figure 15-4 is a diagram of a nephron. Add colored arrows on the figure as
instructed to show the location and direction of the following processes. Add
black arrows at the site of filtrate formation; **red arrows** at the major site of
amino acid and glucose reabsorption; **green arrows** at the sites most
responsive to action of ADH (show direction of water movement); **yellow
arrows** at the sites most responsive to the action of aldosterone (show
direction of Na$^+$ movement); and **blue arrows** at the site of tubular secretion.
Then, label the proximal convoluted tubule (PCT), distal convoluted tubule
(DCT), loop of Henle, Bowman's capsule, and glomerulus on the figure.

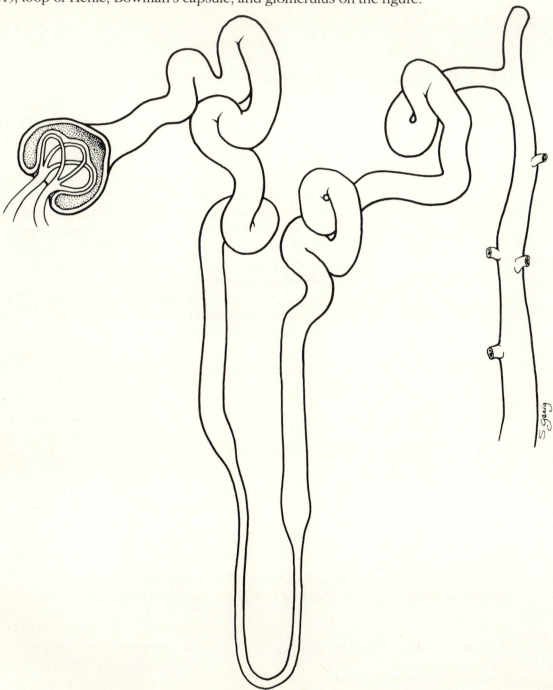

Figure 15-4

18. Circle the term that does not belong in each of the following groupings. NOTE:
    ECF = Extracellular fluid compartment; ICF = Intracellular fluid compartment

   1. Female adult   Male adult   About 50% water   Less muscle

   2. Obese adult   Lean adult   Less body water   More adipose tissue

   3. ECF   Interstitial fluid   ICF   Plasma

   4. Electric charge   Nonelectrolyte   Ions   Conducts a current

   5. ↓ H$^+$ secretion   ↓ Blood pH   ↑ H$^+$ secretion   ↓ K$^+$ secretion

   6. ↑ Water output   ↓ Na$^+$ concentration   ↑ ADH   ↓ ADH

   7. Aldosterone   ↑ Na$^+$ reabsorption   ↑ K$^+$ reabsorption   ↑ BP

   8. ADH   ↓ BP   ↑ Blood volume   ↑ Water reabsorption

Developmental Aspects of the Urinary System

19. Complete the following statements by inserting your responses in the answer blanks.

_____ 1.
_____ 2.
_____ 3.
_____ 4.
_____ 5.
_____ 6.
_____ 7.
_____ 8.
_____ 9.
_____ 10.
_____ 11.
_____ 12.
_____ 13.
_____ 14.

Three separate sets of renal tubules develop in the embryo; however, embryonic nitrogenous wastes are actually disposed of by the __(1)__. A congenital condition typified by blisterlike sacs in the kidneys is __(2)__. __(3)__ is a congenital condition seen in __(4)__, when the urethral opening is located ventrally on the penis. A newborn baby voids frequently, which reflects its small __(5)__. Daytime control of the voluntary urethral sphincter is usually achieved by approximately __(6)__ months. Urinary tract infections are fairly common and not usually severe with proper medical treatment. A particularly problematic condition, called __(7)__, may result later in life as a sequel to childhood streptococcus infection. In this disease, the renal filters become clogged with __(8)__ complexes, urine output decreases, and __(9)__ and __(10)__ begin to appear in the urine. In old age, progressive __(11)__ of the renal blood vessels results in the death of __(12)__ cells. The loss of bladder tone leads to __(13)__ and __(14)__ and is particularly troublesome to elderly people.

The Incredible Journey:
A Visualization Exercise
for the Urinary System

You see the kidney looming brownish red through the artery wall

20. Where necessary, complete statements by inserting the missing word(s) in the answer blanks.

_____ 1.

_____ 2.

_____ 3.

_____ 4.

_____ 5.

_____ 6.

_____ 7.

_____ 8.

_____ 9.

_____ 10.

_____ 11.

_____ 12.

_____ 13.

_____ 14.

_____ 15.

_____ 16.

_____ 17.

_____ 18.

For your journey through the urinary system, you must be made small enough to filter through the filtration membrane from the bloodstream into a renal __(1)__ . You will be injected into the subclavian vein and must pass through the heart before entering the arterial circulation. As you travel through the systematic circulation, you have at least 2 minutes to relax before reaching the __(2)__ artery, feeding a kidney. You see the kidney looming brownish red through the artery wall. Once you have entered the kidney, the blood vessel conduits become increasingly smaller until you finally reach the __(3)__ arteriole, feeding into the filtering device, or __(4)__ . Once in the filter, you maneuver yourself so that you are directly in front of a pore. Within a fraction of a second, you are swept across the filtration membrane into the __(5)__ part of the renal tubule. Drifting along, you lower the specimen cup to gather your first filtrate sample for testing. You study the readout from the sample and note that it is very similar in composition to __(6)__ with one exception: there are essentially no __(7)__ . Your next sample doesn't have to be collected until you reach the "hairpin," or, using the proper terminology, the __(8)__ part of the tubule. As you continue your journey, you notice that the tubule cells have dense fingerlike projections extending from their surfaces into the lumen of the tubule. These are __(9)__ , which increase the surface area of tubules because this portion of the tubule is very active in the process of __(10)__ . Soon you collect your second sample, and then later, in the distal convoluted tubule, your third sample. When you read the computer's summary of the third sample, you make the following notes in your register.

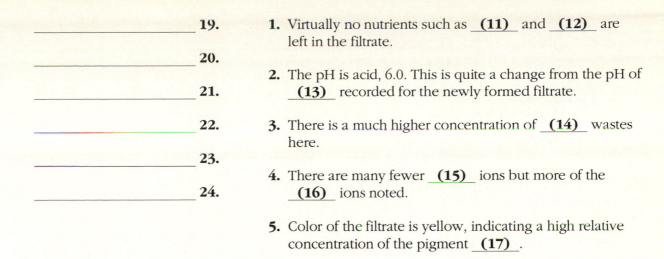

_____ 19.

_____ 20.

_____ 21.

_____ 22.

_____ 23.

_____ 24.

1. Virtually no nutrients such as __(11)__ and __(12)__ are left in the filtrate.

2. The pH is acid, 6.0. This is quite a change from the pH of __(13)__ recorded for the newly formed filtrate.

3. There is a much higher concentration of __(14)__ wastes here.

4. There are many fewer __(15)__ ions but more of the __(16)__ ions noted.

5. Color of the filtrate is yellow, indicating a high relative concentration of the pigment __(17)__ .

Gradually you become aware that you are moving along much more quickly. You see that the water level has dropped dramatically and that the stream is turbulent and rushing. As you notice this, you realize that the hormone __(18)__ must have been released recently to cause this water drop. You take an abrupt right turn and then drop straight downward. You realize that you must be in a __(19)__ . Within a few seconds, you are in what appears to be a large tranquil sea with an ebbing tide toward a darkened area at the far shore. You drift toward the darkened area, confident that you are in the kidney __(20)__ . As you reach and enter the dark tubelike structure seen from the opposite shore, your progress becomes rhythmic—something like being squeezed through a sausage skin. Then you realize that your progress is being regulated by the process of __(21)__ . Suddenly, you free-fall and land in the previously stored __(22)__ in the bladder, where the air is very close. Soon the walls of the bladder begin to gyrate, and you realize you are witnessing a __(23)__ reflex. In a moment, you are propelled out of the bladder and through the __(24)__ to exit from your host.

Reproductive System

The biologic function of the reproductive system is to provide a means for producing offspring. The essential organs are those producing the germ cells (testes in males and ovaries in females). The male manufactures sperm and delivers them to the female's reproductive tract; the female, in turn, produces eggs. If the time is suitable, the fusion of egg and sperm produces a fertilized egg, which is the first cell of the new individual. Once fertilization has occurred, the female uterus protects and nurtures the developing embryo.

In this chapter, student activities concern the structures of the male and female reproductive systems, germ cell formation, the menstrual cycle, and embryonic development.

Male Reproductive System

1. Using the following terms, trace the pathway of sperm from the testis to the urethra: rete testis, epididymis, seminiferous tubule, ductus deferens. List the terms in the proper order in the spaces provided.

    _____ → _____ → _____ → _____ .

2. Name four of the male secondary sex characteristics. Insert your answers on the lines provided.

    _____

    _____

    _____

    _____

3. Using the key choices, select the terms identified in the following descriptions.
    Insert the appropriate term or corresponding letter in the answer blanks.

    KEY CHOICES:

    A. Cowper's glands   **E.** Penis   **I.** Scrotum

    B. Epididymis   **F.** Prepuce   **J.** Spermatic cord

    C. Ductus deferens   **G.** Prostate gland   **K.** Testes

    D. Glans penis   **H.** Seminal vesicles   **L.** Urethra

    _____ **1.** Organ that delivers semen to the female reproductive tract

    _____ **2.** Site of sperm and testosterone production

    _____ **3.** Passageway for conveying sperm from the epididymis to the
                                      ejaculatory duct

    _____ **4.** Conveys both sperm and urine down the length of the penis

    _____ , _____   **5.** Organs that contribute to the formation of semen

    _____ , _____

    _____ **6.** External skin sac that houses the testes

    _____ **7.** Tubular storage site for sperm; hugs the lateral aspect of the testes

    _____ **8.** Cuff of skin, encircling the glans penis

    _____ **9.** Surrounds the urethra at the base of the bladder; produces a milky
                                      alkaline fluid

    _____ **10.** Produces over half of the seminal fluid

    _____ **11.** Empties a lubricating mucus into the urethra

    _____ **12.** Connective tissue sheath enclosing the ductus deferens, blood
                                      vessels, and nerves.

4. Figure 16-1 is a sagittal view of the male reproductive structures. First, identify
    the following organs on the figure by placing each term at the end of the
    appropriate leader line.

    Cowper's gland   Epididymis   Scrotum

    Ductus deferens   Prepuce   Seminal vesicle

    Glans penis   Prostate gland   Testis

    Ejaculatory duct   Urethra

Next, select different colors for the structures that correspond to the
following descriptions, and color in the coding circles and the corresponding
structures on the figure.

◯ Spongy tissue that is engorged with blood during erection

◯ Portion of the duct system that also serves the urinary system

◯ Structure that provides the ideal temperature conditions for sperm formation

◯ Structure removed in circumcision

◯ Gland whose secretion contains sugar to nourish sperm

◯ Structure cut or cauterized during a vasectomy

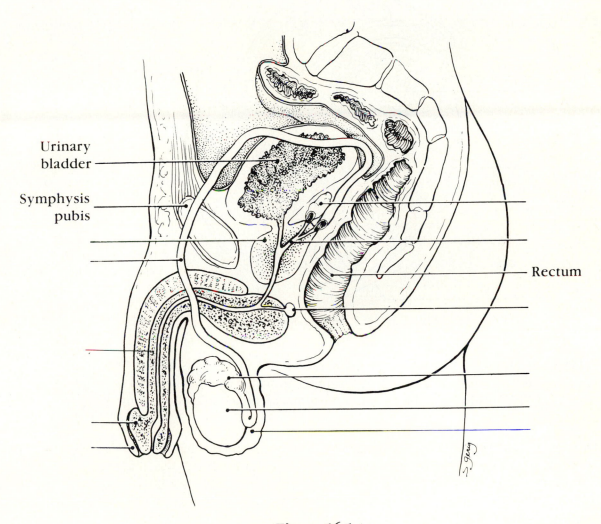

Urinary
bladder

Symphysis
pubis

Rectum

Figure 16-1

5.   Figure 16-2 is a longitudinal section of a testis. First, select different colors for
     the structures that correspond to the following descriptions. Then color the
     coding circles and color and label the corresponding structures on the figure.
     Complete the labeling of the figure by adding the following terms: lobule, rete
     testis, and septum.

     ◯   Site(s) of spermatogenesis

     ◯   Tubular structure in which sperm mature and become motile

     ◯   Fibrous coat protecting the testis

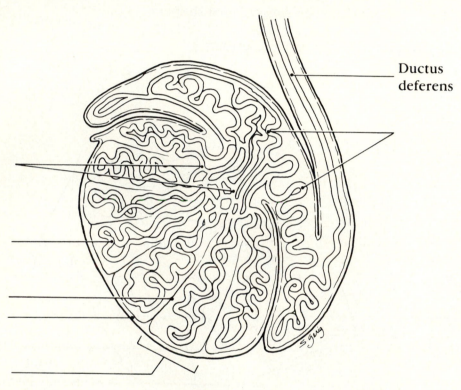

Ductus
deferens

Figure 16-2

6.   This section considers the process of sperm production in the testis. Figure 16-3
     is a cross-sectional view of a seminiferous tubule in which spermatogenesis is
     occurring. First, using key choices, select the terms identified in the following
     descriptions.

KEY CHOICES:

A.   Follicle-stimulating hormone (FSH)   E. ◯   Sperm

B. ◯   Primary spermatocyte   F.   Spermatid

C. ◯   Secondary spermatocyte   G.   Testosterone

D. ◯   Spermatogonium

_____   **1.** Primitive stem cell

_____   **2.** Contain 23 chromosomes

_____

_____   **3.** Product of meiosis I

_____   **4.** Product of meiosis II

_____   **5.** Functional motile gamete

_____   **6.** Hormones necessary for sperm production

_____

Then select different colors for the cell types with color-coding circles listed in the key choices and color in the coding circles and corresponding structures on the figure. In addition, label and color the cells that produce testosterone.

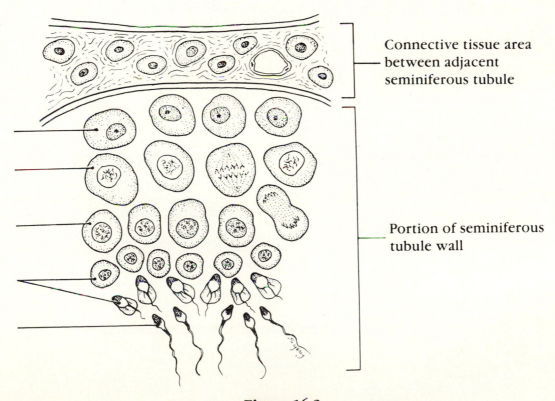

Figure 16-3

7. Figure 16-4 illustrates a single sperm. On the figure, bracket and label the head and the midpiece and circle and label the tail. Select different colors for the structures that correspond to the following descriptions. Color the coding circles and corresponding structures on the figure. Then label the structures, using correct terminology.

○ The DNA-containing area

○ The enzyme-containing sac that aids sperm penetration of the egg

○ Metabolically active organelles that provide ATP to energize sperm movement

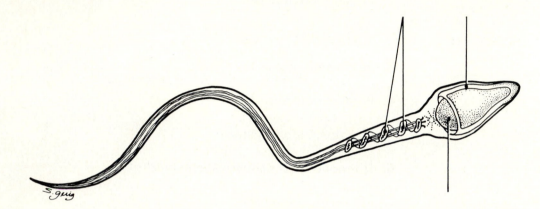

Figure 16-4

8. The following statements refer to events that occur during cellular division. Using key choices, indicate in which type of cellular division the described events occur. Place the correct term or letter response in the answer blanks.

KEY CHOICES:

A. Mitosis   **B.** Meiosis   **C.** Both mitosis and meiosis

_____ **1.** Final product is two daughter cells, each with 46 chromosomes

_____ **2.** Final product is four daughter cells, each with 23 chromosomes

_____ **3.** Involves the phases prophase, metaphase, anaphase, and telophase

_____ **4.** Occurs in all body tissues

_____ **5.** Occurs only in the gonads

_____ **6.** Increases the cell number for growth and repair

_____ **7.** Daughter cells have the same number and types of chromosomes as the mother cell

_____ **8.** Daughter cells are different from the mother cell in their chromosomal makeup

_____ **9.** Chromosomes are replicated before the division process begins

_____ **10.** Provides cells for the reproduction of offspring

_____ **11.** Consists of two consecutive divisions of the nucleus; chromosomes are not replicated before the second division

Female Reproductive System and Cycles

9. Identify the female structures described by inserting your responses in the answer blanks.

_____ **1.** Chamber that houses the developing fetus

_____ **2.** Canal that receives the penis during sexual intercourse

_____ **3.** Usual site of fertilization

_____ **4.** Erects during sexual stimulation

_____ **5.** Duct through which the ovum travels to reach the uterus

_____ **6.** Membrane that partially closes the vaginal canal

_____ **7.** Primary female reproductive organ

_____ **8.** Move to create fluid currents to draw the ovulated egg into the fallopian tube

10. Using the key choices, identify the cell type you would expect to find in the following structures. Insert the correct term or letter response in the answer blanks.

KEY CHOICES:

A. Oogonium   **C.** Secondary oocyte

B. Primary oocyte   **D.** Ovum

_____ **1.** Forming part of the primary follicle in the ovary

_____ **2.** In the uterine tube before fertilization

_____ **3.** In the mature, or graafian, follicle of the ovary

_____ **4.** In the uterine tube shortly after sperm penetration

11. Figure 16-5 is a sagittal view of the female reproductive organs. First, label all structures on the figure provided with leader lines. Then select different colors for the following structures, and use them to color the coding circles and corresponding structures on the figure.

○ Lining of the uterus; endometrium

○ Muscular layer of the uterus, myometrium

○ Pathway along which an egg travels from the time of its release to its implantation

○ Ligament helping to anchor the uterus

○ Structure forming female hormones and gametes

○ Homologue of the male scrotum

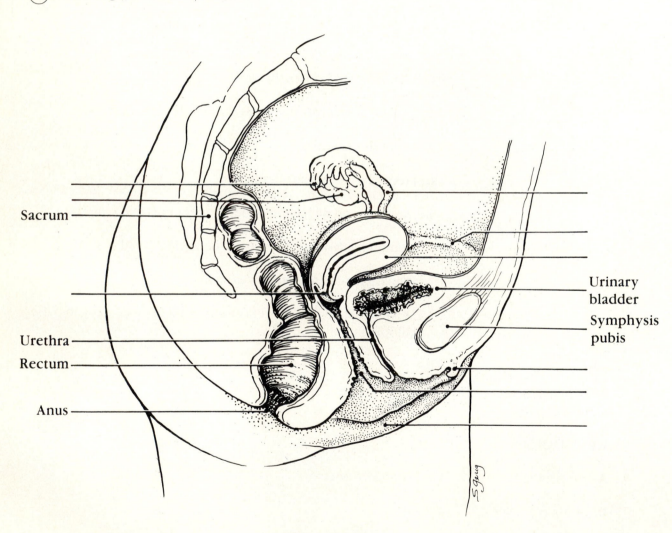

Sacrum

Urethra
Rectum

Anus

Urinary
bladder

Symphysis
pubis

Figure 16-5

12. Figure 16-6 is a ventral view of the female external genitalia. Label the clitoris, labia minora, urethral orifice, hymen, mons pubis, and vaginal orifice on the figure. These structures are indicated with leader lines. Then color the homologue of the male penis blue, color the membrane that partially obstructs the vagina yellow, and color the distal end of the birth canal red.

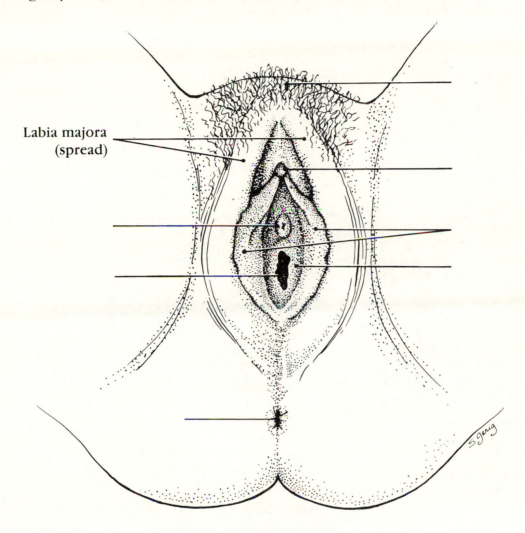

Labia majora
(spread)

Figure 16-6

13. Figure 16-7 is a sectional view of the ovary. First, identify all structures indicated with leader lines on the figure. Second, select different colors for the following structures, and use them to color the coding circles and corresponding structures on the figure.

   ◯ Cells that produce estrogen

   ◯ Glandular structure that produces progesterone

   ◯ All oocytes

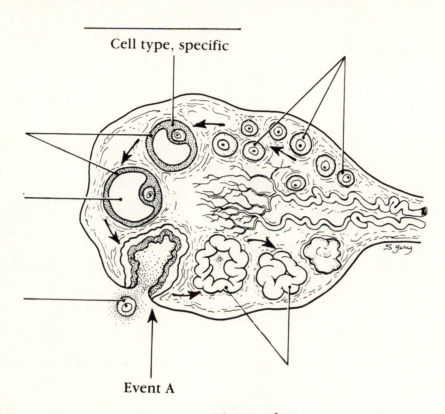

_____
Cell type, specific

Event A

Figure 16-7

Third, in the space provided, name the event depicted as "Event A" on the figure. _____

Fourth, answer the following questions by inserting your answers in the spaces provided.

1. Are there any oogonia in a mature female's ovary? _____

2. Into what area is the ovulated cell released? _____

3. When is a mature ovum (egg) produced in humans? _____

4. What structure in the ovary becomes a corpus luteum? _____

5. What are the four final cell types produced by oogenesis in the female? (Name the cell type and number of each.) _____

6. How does this compare with the final product of spermatogenesis in males? _____

   _____

7. What happens to the tiny cells nearly devoid of cytoplasm ultimately produced during

   oogenesis? _____

8. Why? _____

9. What name is given to the period of a woman's life when her ovaries begin to become

   nonfunctional? _____

14. What is the significance of the fact that the uterine tubes are not structurally
    continuous with the ovaries? Address this question from both reproductive and
    health aspects.

    _____

    _____

    _____

15. The following statements deal with anterior pituitary and ovarian hormonal
    interrelationships. Name the hormone(s) described in each statement. Place
    your answers in the answer blanks.

    _____　　**1.** Promotes growth of ovarian follicles and production of estrogen

    _____　　**2.** Triggers ovulation

    _____　　**3.** Inhibit follicle-stimulating hormone (FSH) release by the anterior
    　　　　　　　　　　　　　　　pituitary

    _____

    _____　　**4.** Stimulates luteinizing hormone (LH) release by the anterior
    　　　　　　　　　　　　　　　pituitary

    _____　　**5.** Converts the ruptured follicle into a corpus luteum and causes it to
    　　　　　　　　　　　　　　　produce progesterone and estrogen

    _____　　**6.** Maintains the hormonal production of the corpus luteum

16. Name four of the secondary sex characteristics of females. Place your answers
    in the spaces provided.

    _____　　　_____

    _____　　　_____

17. Use the key choices to identify the ovarian hormone(s) responsible for the following events. Insert the correct term(s) or letter(s) in the answer blanks.

KEY CHOICES:

A. Estrogen   B. Progesterone

_____   1. Lack of this (these) causes the blood vessels to kink and the endometrium to slough off (menses)

_____   2. Causes the endometrial glands to begin the secretion of nutrients

_____   3. The endometrium is repaired and grows thick and velvety

_____   4. Maintains the myometrium in an inactive state if implantation of an embryo has occurred

_____   5. Glands are formed in the endometrium

_____   6. Responsible for the secondary sex characteristics of females

18. The following exercise refers to Figure 16-8, A-D.

On Figure 16-8, A, the blood levels of two gonadotropic hormones (FSH and LH) of the anterior pituitary are indicated. Identify each hormone by appropriately labeling the blood level lines on the figure. Then select different colors for each of the blood level lines and color them in on the figure.

On Figure 16-8, B, identify the blood level lines for the ovarian hormones, estrogens and progesterone. Then select different colors for each blood level line, and color them in on the figure.

On Figure 16-8, C, select different colors for the following structures and use them to color in the coding circles and corresponding structures in the figure.

◯ Primary follicle   ◯ Growing follicle

◯ Graafian follicle   ◯ Corpus luteum

◯ Ovulating follicle

On Figure 16-8, D, identify the endometrial changes occurring during the menstrual cycle by color-coding and coloring the areas depicting the three phases of that cycle.

◯ Secretory stage   ◯ Menses   ◯ Proliferative stage

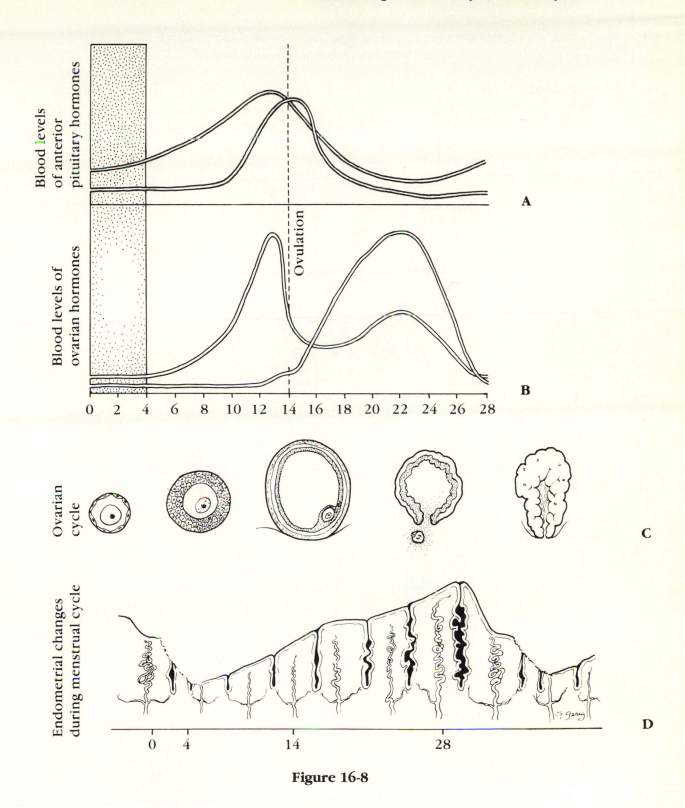

Figure 16-8

19. Figure 16-9 is a sagittal section of a breast. First, use the following terms to correctly label all structures provided with leader lines on the figure.

Alveolar glands   Areola   Lactiferous duct   Nipple

Then color the structures that produce milk blue and color the fatty tissue of the breast yellow.

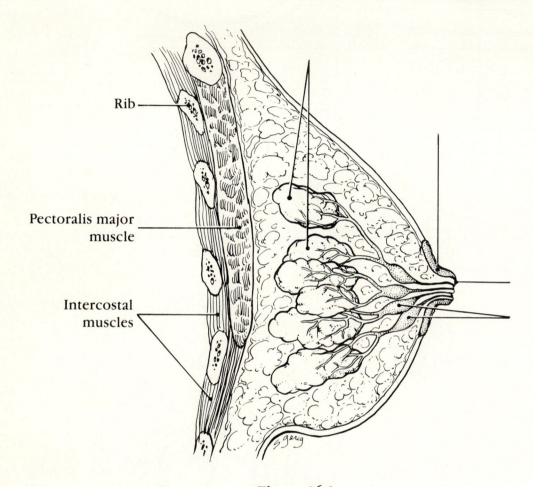

Rib

Pectoralis major muscle

Intercostal muscles

Figure 16-9

Survey of Pregnancy and Embryonic Development

20. Figure 16-10 depicts early embryonic events. Identify the following, referring to the figure. Place your answers in the spaces provided.

   1. Event A _____

   2. Cell A _____

   3. Process B _____

   4. Embryonic structure B_1 _____

   5. Completed process C _____

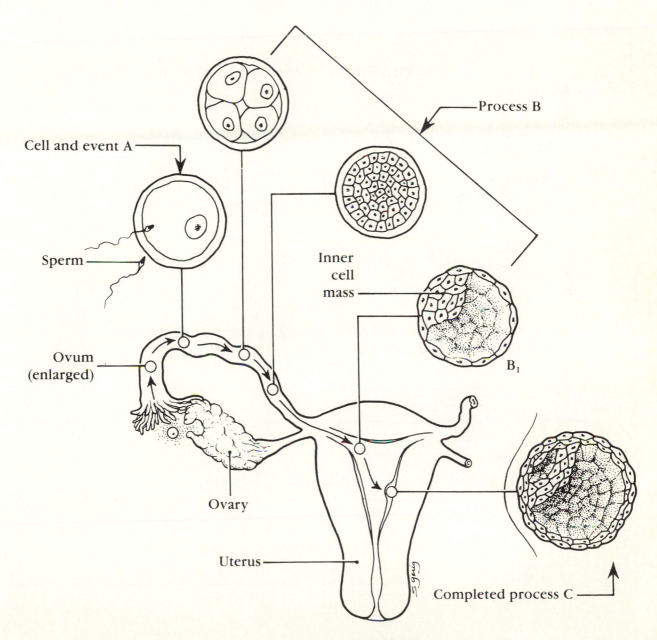

Figure 16-10

21. Using the key choices, select the terms that are identified in the following descriptions. Insert the correct term or letter response in the answer blanks.

KEY CHOICES:

A. Amnion   **D.** Fertilization   **G.** Placenta

B. Chorionic villi   **E.** Fetus   **H.** Umbilical cord

C. Endometrium   **F.** Inner cell mass   **I.** Zygote

_____ **1.** Part of the blastocyst that becomes the embryonic body

_____ **2.** Secretes estrogen and progesterone to maintain the pregnancy

_____ **3.** Cooperate to form the placenta

_____

_____ **4.** Fluid-filled sac, surrounding the developing embryo/fetus

_____ **5.** Attaches the embryo to the placenta

_____ **6.** Fingerlike projections of the blastocyst

_____ **7.** The embryo after 8 weeks

_____ **8.** The organ that delivers nutrients to and disposes of wastes for the fetus

_____ **9.** Event leading to combination of ovum and sperm "genes"

_____ **10.** The fertilized egg

22. Explain why the corpus luteum does not stop producing its hormones (estrogen and progesterone) when fertilization has occurred.

_____

_____

_____

23. What two hormones are essential to precipitate labor in humans?

_____

24. **1.** What hormone is responsible for milk production? _____

2. For milk ejection? _____

25. A pregnant woman undergoes numerous changes during her pregnancy—
    anatomical, metabolic, and physiological. Several such possibilities are listed
    below. Check (✓) all that are commonly experienced during pregnancy.

    _____ **1.** Glucose sparing

    _____ **2.** Breasts decline in size

    _____ **3.** Pelvic ligaments are relaxed by
    relaxin

    _____ **4.** Vital capacity decreases

    _____ **5.** Lordosis

    _____ **6.** Blood pressure and pulse rates
    decline

    _____ **7.** Metabolic rate declines

    _____ **8.** Increased mobility of GI tract

    _____ **9.** Blood volume and cardiac output
    increase

    _____ **10.** Nausea, heartburn, constipation

    _____ **11.** Dyspnea may occur

    _____ **12.** Urgency and stress incontinence

    _____ **13.** Diaphragm descent is impaired

26. What are Braxton Hicks contractions, and why do they occur?

    _____

    _____

    _____

    _____

27. Name the three phases of parturition, and briefly describe each phase.

    1. _____

    _____

    2. _____

    _____

    3. _____

    _____

28. The very simple flow chart in Figure 16-11 illustrates the sequence of events that occur during labor. Complete the flow chart by filling in the missing terms in the boxes. Use color as desired.

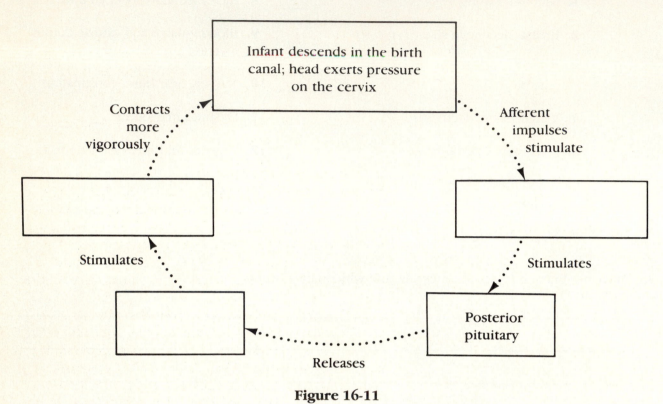

Figure 16-11

29. How long will the cycle illustrated in Figure 16-11 continue to occur?

_____

30. Labor is an example of a positive feedback mechanism. What does that mean?

_____

_____

_____

Developmental Aspects of the Reproductive System

31. Complete the following statements by inserting your responses in the answer blanks.

1. _____
2. _____
3. _____
4. _____
5. _____
6. _____
7. _____
8. _____
9. _____
10. _____
11. _____
12. _____
13. _____
14. _____
15. _____
16. _____
17. _____
18. _____
19. _____
20. _____
21. _____
22. _____

A male embryo has **(1)** sex chromosomes, whereas a female has **(2)** . During early development, the reproductive structures of both sexes are identical, but by the eighth week, **(3)** begin to form if testosterone is present. In the absence of testosterone, the female **(4)** forms. The testes of a male fetus descend to the scrotum shortly before birth. If this does not occur, the resulting condition is called **(5)** .

The most common problem affecting the reproductive organs of women are infections, particularly **(6)** , **(7)** , and **(8)** . When the entire pelvis is inflamed, the condition is called **(9)** . Most male problems involve inflammations, resulting from **(10)** . A leading cause of cancer death in adult women is cancer of the **(11)** ; the second most common female reproductive system cancer is cancer of the **(12)** . Thus a yearly **(13)** is a very important preventive measure for early detection of this latter cancer type. The cessation of ovulation in an aging woman is called **(14)** . Intense vasodilation of blood vessels in the skin lead to uncomfortable **(15)** . Additionally, bone mass **(16)** and blood levels of cholesterol **(17)** when levels of the hormone **(18)** wane. In contrast, healthy men are able to father children well into their eighth decade of life. Postmenopausal women are particularly susceptible to **(19)** inflammations. The single most common problem of elderly men involves the enlargement of the **(20)** , which interferes with the functioning of both the **(21)** and **(22)** systems.

The Incredible Journey:
A Visualization Exercise
for the Reproductive System

*. . . you hear a piercing sound coming from the
almond-shaped organ as its wall ruptures.*

32. Where necessary, complete statements by inserting the missing word(s) in the
answer blanks.

_____ 1.

_____ 2.

_____ 3.

_____ 4.

_____ 5.

_____ 6.

_____ 7.

_____ 8.

_____ 9.

_____ 10.

_____ 11.

_____ 12.

_____ 13.

_____ 14.

_____ 15.

_____ 16.

_____ 17.

_____ 18.

_____ 19.

_____ 20.

This is your final journey. You are introduced to a hostess
this time, who has agreed to have her cycles speeded up by
megahormone therapy so that all of your observations can
be completed in less than a day. Your instructions are to
observe and document as many events of the two female
cycles as possible.

You are miniaturized to enter your hostess through a tiny
incision in her abdominal wall (this procedure is called a
laparotomy, or, more commonly, "belly button surgery") and
end up in her peritoneal cavity. You land on a large and
pear-shaped organ in the abdominal cavity midline,
the __(1)__ . You survey the surroundings and begin to make
organ identifications and notes of your observations.
Laterally and way above you on each side is an
almond-shaped __(2)__ , which is suspended by a ligament
and almost touched by "featherduster-like" projections of a
tube snaking across the abdominal cavity toward the
almond-shaped organs. The projections appear to be almost
still, which is puzzling because you thought that they were
the __(3)__ , or fingerlike projections, of the ovarian tubes,
which are supposed to be in motion. You walk toward the
end of one of the ovarian tubes to take a better look. As you
study the ends of the ovarian tube more closely, you
discover that the featherlike projections are now moving
more rapidly, as if they are trying to coax something into the
ovarian tube. Then you spot a reddened area on the
almond-shaped organ, which seems to be enlarging even as
you watch. As you continue to observe the area, you waft
gently up and down in the peritoneal fluid. Suddenly you
feel a gentle but insistent sucking current, drawing you
slowly toward the ovarian tube. You look upward and see
that the reddened area now looks like an angry boil, and the
ovarian tube projections are gyrating and waving frantically.
You realize that you are about to witness __(4)__ . You try to
get still closer to the opening of the ovarian tube, when you
hear a piercing sound coming from the almond-shaped
organ as its wall ruptures. Then you see a ball-like structure,

with a "halo" of tiny cells enclosing it, being drawn into the ovarian tube. You have just seen the
__(5)__ , surrounded by its capsule of __(6)__ cells, entering the ovarian tube. You hurry into the ovarian
tube behind it and, holding onto one of the tiny cells, follow it to the uterus. The cell mass that you
have attached to has no way of propelling itself, yet you are being squeezed along toward the uterus
by a process called __(7)__ . You also notice that there are __(8)__ , or tiny hairlike projections of the tubule cells, that are all waving in the same direction as you are moving.

Nothing seems to change as you are carried along until finally you are startled by a deafening noise.
Suddenly there are thousands of tadpole-like __(9)__ swarming all around you and the sphere of cells.
Their heads seem to explode as their __(10)__ break and liberate digestive enzymes. The cell mass now
has hundreds of openings in it, and some of the small cells are beginning to fall away. As you peer
through the rather transparent cell "halo" you see that one of the tadpole-like structures has pene-
trated the large central cell. Chromosomes then appear, and that cell begins to divide. You have just
witnessed the second __(11)__ division. The products of this division are one large cell, the __(12)__ ,
and one very tiny cell, the __(13)__ , which is now being ejected. This cell will soon be __(14)__ because
it has essentially no cytoplasm or food reserves. As you continue to watch, the sperm nucleus and that
of the large central cell fuse, an event called __(15)__ . You note that the new cell just formed by this fusion is called a __(16)__ , the first cell of the embryonic body.

As you continue to move along the fallopian tube, the central cell divides so fast that no cell growth occurs between the divisions. Thus the number of cells forming the embryonic body increases, but the
cells become smaller and smaller. This embryonic division process is called __(17)__ .

Finally, the uterine chamber looms before you. As you drift into its cavity, you scrutinize its lining, the
__(18)__ . You notice that it is thick and velvety in appearance and that the fluids you are drifting in are
slightly sweet. The embryo makes its first contact with the lining, detaches, and then makes a second
contact at a slightly more inferior location. This time it sticks, and as you watch, the lining of the organ
begins to erode away. The embryo is obviously beginning to burrow into the rich cushiony lining, and
you realize that __(19)__ is occurring.

You now leave the embryo and propel yourself well away from it. As you float in the cavity fluids, you
watch the embryo disappear from sight beneath the lining. Then you continue to travel downward
through your hostess's reproductive tract, exiting her body at the external opening of the __(20)__ .

Notes

Answers

Chapter 1

1. **1:** D or physiology.   **2:** A or anatomy.   **3:** B or homeostasis.   **4:** C or metabolism.
2. Physiological study: C, D, E, F, G, H, J.   Anatomical study: A, B, I, K, L, M.
3. Cells, tissues, organs, organ systems.
4. **1:** Atom.   **2:** Epithelium.   **3:** Heart.   **4:** Digestive system.
5. **1:** J or urinary.   **2:** C or endocrine.   **3:** I or skeletal.   **4:** A or circulatory.   **5:** D or integumentary.   **6:** A or circulatory.   **7:** B or digestive.   **8:** H or respiratory.   **9:** A or circulatory.   **10:** E or muscular.   **11:** J or urinary.   **12:** G or reproductive.   **13:** C or endocrine.   **14:** D or integumentary.
6. **1:** A or circulatory.   **2:** C or endocrine.   **3:** J or urinary.   **4:** G or reproductive.   **5:** B or digestive.   **6:** I or skeletal.   **7:** F or nervous.
7.

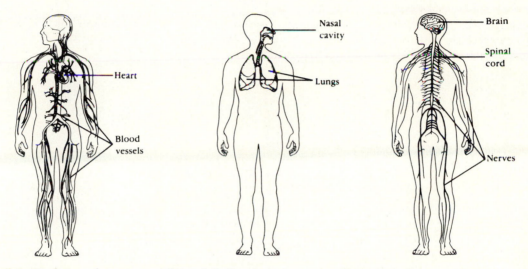

Figure 1-1: Cardiovascular system   **Figure 1-2:** Respiratory system   **Figure 1-3:** Nervous system

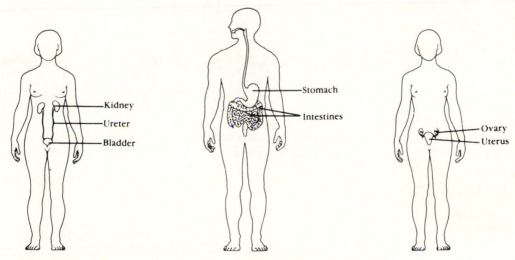

Figure 1-4: Urinary system   **Figure 1-5:** Digestive system   **Figure 1-6:** Reproductive system

8. **1:** D or maintenance of boundaries.   **2:** H or reproduction.   **3:** C or growth.   **4:** A or digestion.   **5:** B or excretion.   **6:** G or responsiveness.   **7:** F or movement.   **8:** E or metabolism.   **9:** D or maintenance of boundaries.
9. **1:** C or nutrients.   **2:** B or atmospheric pressure.   **3:** E or water.   **4:** D or oxygen.   **5:** E or water.   **6:** A or appropriate body temperature.
10. **1:** Receptor.   **2:** Control center.   **3:** Afferent.   **4:** Control Center.   **5:** Effector.   **6:** Efferent.   **7:** Negative.   **8:** Positive.   **9:** Negative.
11. **1:** Ventral.   **2:** Dorsal.   **3:** Dorsal.

12.   **1:** Distal.   **2:** Cubital.   **3:** Brachial.   **4:** Left upper quadrant.
13.   **Figure 1-7:**

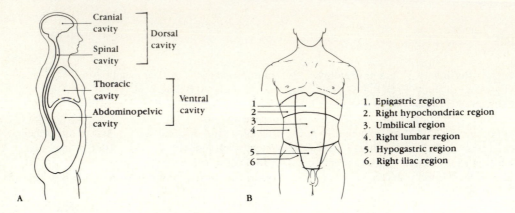

14.   **1:** C or axillary.   **2:** H or femoral.   **3:** I or gluteal.   **4:** G or cervical.   **5:** O or umbilical.   **6:** N or pubic.
       7: B or antecubital.   **8:** L or occipital.   **9:** J or inguinal.   **10:** K or lumbar.   **11:** E or buccal.
15.   **Figure 1-8: Section A:** Midsagittal.   **Section B:** Transverse.

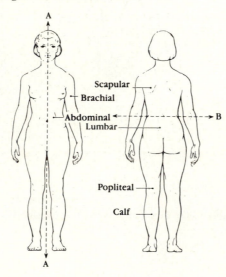

16.   **1:** G or ventral, D or pelvic.   **2:** G or ventral, F or thoracic.   **3:** C or dorsal, B or cranial.   **4:** G or ventral, A or
       abdominal.   **5:** G or ventral, A or abdominal.
17.   **1:** A or anterior.   **2:** G or posterior.   **3:** J or superior.   **4:** J or superior.   **5:** E or lateral.   **6:** A or anterior.
       7: F or medial.   **8:** H or proximal.   **9:** B or distal.   **10:** G or posterior.   **11:** J or superior.   **12:** I or sagittal.
       13: K or transverse.   **14:** C or frontal.   **15:** K or transverse.
18.   **1:** A or abdominopelvic.   **2:** A or abdominopelvic.   **3:** A or abdominopelvic.   **4:** A or abdominopelvic.
       5: A or abdominopelvic.   **6:** C or spinal.   **7:** A or abdominopelvic.   **8:** D or thoracic.   **9:** D or thoracic.
       10: B or cranial.   **11:** A or abdominopelvic.   **12:** A or abdominopelvic.
19.   **1:** 2, 3, 7, 11, 12.   **2:** 2, 3.   **3:** 2.   **4:** 1, 2, 3, 5.   **5:** 2, 3.

Chapter 2

1.   pH is the symbol of hydrogen ion concentration; a measure of the relative acidity or basicity of a solution or substance.
2.   **1:** Milk of magnesia.   **2:** Urine (pH 6.2).   **3:** $NaHCO_2$.   **4:** Organic.   **5:** pH 4.
3.   **1:** E or ion.   **2:** F or matter.   **3:** C or element.   **4:** B or electrons.   **5:** B or electrons.   **6:** D or energy.
       7: A or atom.   **8:** G or molecule.   **9:** I or protons.   **10:** J or valence.   **11:** I or protons.   **12:** H or neutrons.
4.   **1:** O.   **2:** C.   **3:** K.   **4:** I.   **5:** H.   **6:** N.   **7:** Ca.   **8:** Na.   **9:** P.   **10:** Mg.   **11:** Cl.   **12:** Fe.
5.   **1:** C or synthesis.   **2:** B or exchange.   **3:** A or decomposition.
6.   **Figure 2-1:** The nucleus is the innermost circle containing 6P and 6N; the electrons are indicated by the small circles in
       the orbits.   **1:** Atomic number is 6.   **2:** Atomic mass is 12 amu.   **3:** Carbon.   **4:** Isotope.   **5:** Chemically active.
       6: Four electrons.   **7:** Covalent because it would be very difficult to gain or lose four electrons.

7. H_2O_2 is one molecule of hydrogen peroxide (a compound). 2 OH⁻ represents two hydroxide ions.
8. **Figure 2-2:**

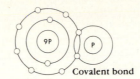

Ionic bond   Covalent bond

9. **1-3:** A or acid(s), B or base(s), and D or salt.   **4:** B or base(s).   **5:** A or acid(s).   **6:** D or salt.   **7:** D or salt.   **8.** A or acid(s).   **9:** C or buffer.
10. **1:** *T.*   **2:** Colloid.   **3:** Acidic.   **4:** *T.*   **5:** Covalent.   **6:** Suspension.   **7:** Cation.   **8:** Protons.   **9:** Carbon. **10:** Solvent.   **11:** More.   **12:** *T.*   **13:** Radioactive.
11. **1:** H_2CO_3.   **2:** H^+ and HCO_3^-.   **3:** The ions should be circled.   **4:** An additional arrow going to the left should be added between H_2CO_3 and H^+.
12. *X* nucleic acids, fats, proteins, and glucose.
13. **1:** H or monosaccharides.   **2 and 3:** D or fatty acids and E or glycerol.   **4:** A or amino acids.   **5:** G or nucleotides. **6:** I or proteins.   **7:** B or carbohydrates.   **8:** C or fats.   **9:** H or or monosaccharides (B or carbohydrates). **10:** C or fats.   **11 and 12:** G or nucleotides and I or proteins.
14. **1:** B or collagen, H or keratin.   **2:** D or enzyme, F or hemoglobin, some of G or hormones.   **3:** D or enzyme. **4:** L or starch.   **5:** E or glycogen.   **6:** C or DNA.   **7:** A or cholesterol.   **8:** I or lactose, J or maltose.
15. **Figure 2-3, A:** Monosaccharide.   **B:** Protein.   **C:** Nucleotide.   **D:** Fat.   **E:** Polysaccharide.
16. **1:** Glucose.   **2:** Ribose.   **3:** Glycogen.   **4:** Glycerol.   **5:** Glucose.   **6:** Hydrolysis.
17. Unnamed nitrogen bases: Thymine (T) and Guanine (G). **1:** Hydrogen bonds.   **2:** Double helix.   **3:** 12. **4:** Complementary.
    Figure 2-4:

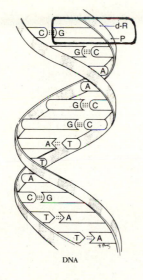

Note that the stippled parts of the backbones represent phosphate units (P) while the unaltered (white) parts of the backbones that are attached to the bases are deoxyribose sugar (d-R) units.

DNA

18. **1:** *T.*   **2:** Neutral fats.   **3:** *T.*   **4:** Polar.   **5:** *T.*   **6:** ATP.   **7:** *T.*   **8:** O.
19. **1:** Negative.   **2:** Positive.   **3:** Hydrogen bonds.   **4:** Red blood cells.   **5:** Protein.   **6:** Amino acids.   **7:** Peptide. **8:** H^+ and OH⁻.   **9:** Hydrolysis or digestion.   **10:** Enzyme.   **11:** Glucose.   **12:** Glycogen.   **13:** Synthesis. **14:** H_2O.   **15:** Increase.

Chapter 3

1. **1-4:** (in any order): Carbon; Oxygen; Nitrogen; Hydrogen.   **5:** Water.   **6:** Calcium.   **7:** Iron.   **8-12:** (five of the following, in any order): Metabolism; Reproduction; Irritability; Mobility; Ability to grow; ability to digest foods; ability to excrete waste.   **13-15:** (three of the following, in any order): Cubelike; Tilelike; Disk; Round spheres; Branching; Cylindrical.   **16:** Tissue (interstitial) fluid.   **17:** Squamous epithelial.

2.

Cell structure	Location	Function
Plasma membrane	External boundary of the cell	Confines cell contents; regulates entry and exit of materials
Lysosome	Scattered in cytoplasm	Digests ingested materials and worn-out organelles
Mitochondria	Scattered throughout the cell	Controls release of energy from foods; forms ATP
Microvilli	Projections of the cell membrane	Increase the membrane surface area
Golgi apparatus	Near the nucleus (in the cytoplasm	Packages proteins to be exported from the cell
Nucleus	(Usually) center of the cell	Storehouse of genetic information; directs cellular activities, including division
Centrioles	Two rod-shaped bodies near the nucleus	"Spin" the mitotic spindle
Nucleolus	Dark spherical body in the nucleus	Storehouse/assembly site for ribosomes
Smooth ER	In the cytoplasm	Site of steroid synthesis
Rough ER	In the cytoplasm	Transports proteins (made on its ribosomes) to other sites in the cell
Ribosomes	Attached to membrane systems or scattered in the cytoplasm	Synthesize proteins
Chromatin	Dispersed in the nucleus	Contains genetic material (DNA); coils during mitosis
Inclusions	Dispersed in the cytoplasm	Provides nutrients; represents cell waste, stored products, etc.

3.   **Figure 3-1:**

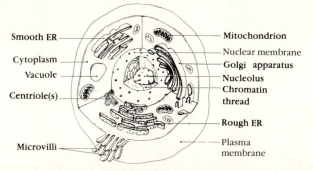

4.   **1:** Q or protein.   **2:** G or double helix.   **3:** P or phosphate.   **4:** U or sugar.   **5:** C or bases.   **6:** B or amino acids.
7: E or complementary.   **8:** F or cytosine.   **9:** X or thymine.   **10:** T or ribosome.   **11:** V or structural proteins.
12: I or functional proteins.   **13:** H or enzymes.   **14:** R or replication.   **15:** N or nucleotides.   **16:** W or template
or model.   **17:** M or new.   **18:** O or old.   **19:** J or genes.   **20:** K or growth.   **21:** S or repair.
5.   **1:** B, C, E.   **2:** A, D, E.
6.   **1:** A.   **2:** B.   **3:** C.   **4:** A.
7.   **Figure 3-2. 1:** A; Crenated.   **2:** B; The same solute concentration inside and outside the cell.   **3:** C; They are bursting
(lysis); Water is moving by osmosis from its site of higher concentration (cell exterior) into the cell where it is in lower
concentration, causing the cells to swell.

8. **1:** D or phagocytosis, E or pinocytosis, F or solute pumping.   **2:** A or diffusion, dialysis, B or diffusion, osmosis.
    3: C or filtration.   **4:** A or diffusion, dialysis, B or diffusion, osmosis.   **5:** F or solute pumping.   **6:** D or
    phagocytosis, E or pinocytosis.   **7:** B or diffusion, osmosis.   **8:** F or solute pumping.   **9:** D or phagocytosis.
    10: A or diffusion, dialysis.

9. **Figure 3-3:** Oxygen and fats will move from the cell exterior into the cell by moving passively through the lipid portion;
    the lipid portion is indicated as the major part of the membrane composed of small spheres, each with two "tails."
    Amino acids and glucose also enter the cell from the exterior by attaching to a protein carrier (indicated as large, solid,
    irregularly shaped structures extending part or all of the way though the membrane). Carbon dioxide, like oxygen,
    passes by diffusion through the lipid part of the membrane, but in the opposite direction (i.e., from the cell interior to
    the cell exterior).

10. **Figure 3-4. A:** Prophase.   **B:** Anaphase.   **C:** Telophase.   **D:** Metaphase.

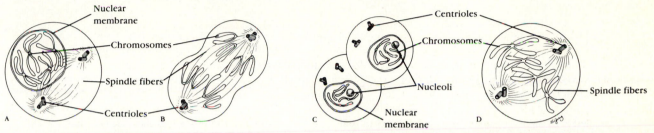

11. **1:** C or prophase.   **2:** A or anaphase.   **3:** D or telophase.   **4:** D or telophase.   **5:** B or metaphase.   **6:** C or
    prophase.   **7:** C or prophase.   **8:** E or none of these.   **9:** C or prophase.   **10:** C or prophase.   **11:** E or none of
    these.   **12:** A or anaphase, B or metaphase.   **13:** E or none of these.

12. **1:** Nucleus.   **2:** Cytoplasm.   **3:** Coiled.   **4:** Centromeres.   **5:** Binucleate cell.   **6:** Spindle.   **7:** Interphase.

13. **Figure 3-5, A:** Simple squamous epithelium.   **B:** Simple cuboidal epithelium.   **C:** Cardiac muscle.   **D:** Dense fibrous
    connective tissue.   **E:** Bone.   **F:** Skeletal muscle.   **G:** Nervous tissue.   **H:** Hyaline cartilage.   **I:** Smooth muscle
    tissue.   **J:** Adipose (fat) tissue.   **K:** Stratified squamous epithelium.   **L:** Areolar connective tissue.   The non-cellular
    portions of D, E, H, J, and L are matrix.

14. The neuron has long cytoplasmic extensions that promote its ability to transmit impulses over long distances within the
    body.

15. **1:** B or epithelium.   **2:** C or muscle.   **3:** D or nervous.   **4:** A or connective.   **5:** B or epithelium.   **6:** D or nervous.
    7: C or muscle.   **8:** B or epithelium.   **9:** A or connective.   **10:** A or connective.   **11:** C or muscle.   **12:** A or
    connective.   **13:** D or nervous.

16. **1:** E or stratified squamous.   **2:** B or simple columnar.   **3:** D or simple squamous.   **4:** C or simple cuboidal.
    5: E or stratified squamous.   **6:** F or transitional.   **7:** D or simple squamous.

17. **1:** Skeletal.   **2:** Cardiac; smooth.   **3:** Skeletal; cardiac.   **4:** Smooth (most cardiac)   **5-7:** Skeletal.
    8 and 9: Smooth.   **10:** Cardiac.   **11:** Skeletal.   **12:** Cardiac.   **13:** Skeletal.   **14:** Smooth.   **15:** Cardiac.

18. **1:** Areolar.   **2:** Cell.   **3:** Elastic fibers.   **4:** Bones.   **5:** Nervous.   **6:** Blood.

19. **1:** C or dense fibrous.   **2:** A or adipose.   **3:** C or dense fibrous.   **4:** D or osseous tissue.   **5:** B or areolar.   **6:** F or
    hyaline cartilage.   **7:** A or adipose.   **8:** F or hyaline cartilage.   **9:** D or osseous tissue.

20. **1:** Inflammation.   **2:** Clotting proteins.   **3:** Granulation.   **4:** Regeneration.   **5:** *T.*   **6:** Collagen.   **7:** *T.*

21. **1:** Tissues.   **2:** Growth.   **3:** Nervous.   **4:** Muscle.   **5:** Connective (scar).   **6:** Chemical.   **7:** Physical.   **8:** Genes
    (DNA).   **9:** Connective tissue changes.   **10:** Decreased endocrine system activity.   **11:** Dehydration of body
    tissues.   **12:** Division.   **13:** Benign.   **14:** Malignant.   **15:** Benign.   **16:** Malignant.   **17:** Biopsy.   **18:** Surgical
    removal.   **19:** Hyperplasia.   **20:** Atrophy.

22. **1:** Cytoplasm.   **2:** Nucleus.   **3:** Mitochondrion.   **4:** ATP.   **5:** Ribosomes.   **6:** Rough endoplasmic reticulum.
    7: Pores.   **8:** Chromatin.   **9:** DNA.   **10:** Nucleoli.   **11:** Golgi apparatus.   **12:** Lysosome.

Chapter 4

1. **1:** Heat.   **2:** Subcutaneous.   **3:** Keratin.   **4:** Vitamin D.   **5:** A freckle.   **6:** Elasticity.   **7:** Oxygen (blood flow).
    8: Cyanosis.

2. **Figure 4-1:**

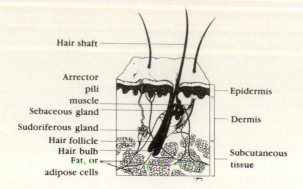

3. **1:** A or stratum corneum, D or stratum lucidum.   **2:** A or stratum corneum, D or stratum lucidum.   **3:** E or papillary layer.   **4:** H or dermis as a whole.   **5:** B or stratum germinativum.   **6:** A or stratum corneum.   **7:** H or dermis as a whole.   **8:** B or stratum germinativum.   **9:** G or epidermis as a whole.

4. **1:** C or third-degree burn.   **2:** B or second-degree burn.   **3:** A or first-degree burn.   **4:** B or second-degree burn.   **5:** C or third-degree burn.   **6:** C or third-degree burn.

5. Infection and water/protein/electrolyte loss.

6. **1:** E or sebaceous glands.   **2:** A or arrector pili.   **3:** G or eccrine sudoriferous glands.   **4:** D or hair follicle.   **5:** F or apocrine sudoriferous glands.   **6:** C or hair.   **7:** B or cutaneous receptors.   **8:** E or sebaceous glands, and F or apocrine sudoriferous glands.   **9:** G or eccrine sudoriferous glands.

7. **1:** Arrector pili.   **2:** Absorption.   **3:** Stratum germinativum.   **4:** Blackheads.   **5:** Eccrine glands.

8. **1:** Melanin.   **2:** Keratin.   **3:** *T.*   **4:** Stratum corneum.   **5:** Shaft.   **6:** Dermis.

9. **Figure 4-2:**

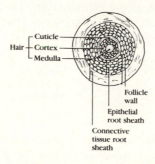

10. The mucous, serous, and cutaneous membranes are all composite membranes composed of an epithelial layer underlaid by a connective tissue layer. A mucous membrane is an epithelial sheet underlaid by a connective tissue layer called the lamina propria. Mucosae line the respiratory, digestive, urinary, and reproductive tracts; functions include protection, lubrication, secretion, and absorption. Serous membranes consist of a layer of simple squamous epithelium resting on a scant layer of fine connective tissue. Serosae line internal ventral body cavities and cover their organs; their function is to produce a lubricating fluid that reduces friction. The cutaneous membrane, or skin, is composed of the epithelial epidermis and the connective tissue dermis. It covers the body exterior and protects deeper body tissues from external insults. The synovial membranes, which line joint cavities of synovial joints, are composed entirely of connective tissue. They function to produce lubrication to decrease friction within the joint cavity.

11. In each case, the visceral layer of the serosa covers the external surface of the organ, and the parietal layer lines the body cavity walls.

    Figure 4-3:

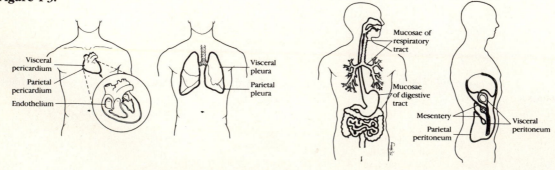

12. **1:** C or dermatitis.   **2:** D or delayed action gene.   **3:** F or milia.   **4:** B or cold intolerance.   **5:** A or acne.   **6:** G or vernix caseosa.   **7:** E or lanugo.

13. Anxiety, chemotherapy or radiation therapy, and fungus diseases (and excessive vitamin A).

14. **1:** Collagen. **2:** Elastin (or elastic). **3:** Dermis. **4:** Phagocyte (macrophage). **5:** Hair follicle. **6:** Epidermis. **7:** Stratum germinativum. **8:** Melanin. **9:** Keratin. **10:** Squamous (stratum corneum) cells.

Chapter 5

1. **1:** P. **2:** P. **3:** D. **4:** D. **5:** P. **6:** D. **7:** P. **8:** P. **9:** P.
2. **1:** S. **2:** F. **3:** L. **4:** L. **5:** F. **6:** L. **7:** L. **8:** F. **9:** I.
3. **1:** C or epiphysis. **2:** A or diaphysis. **3:** C or epiphysis, D or red marrow. **4:** A or diaphysis. **5:** E or yellow marrow cavity. **6:** B or epiphyseal plate.
4. **1:** G or parathyroid hormone. **2:** F or osteocytes. **3:** A or atrophy. **4:** H or stress. **5:** D or osteoblasts. **6:** B or calcitonin. **7:** E or osteoclasts. **8:** C or gravity.
5. **Figure 5-1. 1:** B or lamellae. **2:** C or lacunae. **3:** A or haversian canal. **4:** E or matrix. **5:** D or canaliculi. Matrix is the nonliving part of bone shown as the unlabeled white space on the illustration.

 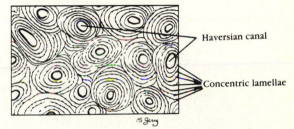

6. **Figure 5-2:** The epiphyseal plate is the white band shown in the center region of the head; the articular cartilage is the white band on the external surface of the head. Red marrow is found within the spongy bone cavity; yellow marrow is found within the cavity of the diaphysis.

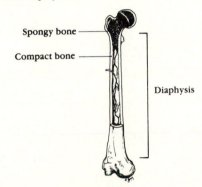

7. **1:** B or frontal. **2:** N or zygomatic. **3:** E or mandible. **4:** G or nasals. **5:** I or palatines. **6:** J or parietals. **7:** H or occipital. **8:** K or sphenoid. **9:** D or lacrimals. **10:** F or maxillae. **11:** A or ethmoid. **12:** L or temporals. **13:** K or sphenoid. **14:** A or ethmoid. **15:** E or mandible. **16:** L or temporals. **17-20:** A or ethmoid, B or frontal, F or maxillae, and K or sphenoid. **21:** H or occipital. **22:** H or occipital. **23:** L or temporals. **24:** M or vomer. **25:** A or ethmoid.
8. **Figure 5-3:**

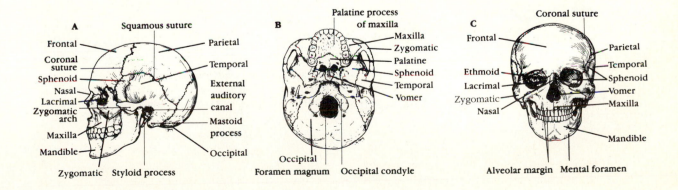

9. **Figure 5-4. 1:** Mucosa-lined air-filled cavities in bone.   **2:** They lighten the skull and serve as resonance chambers for speech.   **3:** Their mucosa is continuous with that of the nasal passages into which they drain.

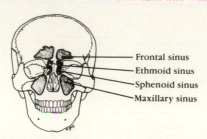

Frontal sinus
Ethmoid sinus
Sphenoid sinus
Maxillary sinus

10. **1:** F or vertebral arch.   **2:** A or body.   **3:** C or spinous process, D or superior articular process, E or transverse process.   **4:** A or body, E or transverse process.   **5:** B or intervertebral foramina.
11. **1:** A or atlas, B or axis, C or cervical vertebra—typical.   **2:** B or axis.   **3:** G or thoracic vertebra.   **4:** F or sacrum.   **5:** E or lumbar vertebra.   **6:** D or coccyx.   **7:** A or atlas.   **8:** C or cervical vertebra—typical.   **9:** G or thoracic vertebra.
12. **1:** Kyphosis.   **2:** Scoliosis.   **3:** Fibrocartilage.   **4:** Springiness or flexibility.
13. **Figure 5-5, A:** Cervical; atlas.   **B:** Cervical.   **C:** Thoracic.   **D:** Lumbar.
14. **Figure 5-6. 1:** Cervical, C_1-C_7.   **2:** Thoracic, T_1-T_{12}.   **3:** Lumbar, L_1-L_5.   **4:** Sacrum, fused.   **5:** Coccyx, fused.   **6:** Atlas, C_1 **7:** Axis, C_2.
15. **1:** Lungs.   **2:** Heart.   **3:** True.   **4:** False.   **5:** Floating.   **6:** Thoracic vertebrae.   **7:** Sternum.   **8:** An inverted cone.
16. **Figure 5-7:** Ribs #1-#7 on each side are true ribs; ribs #8-#12 on each side are false ribs.   17.   **Figure 5-8:** Scapula.

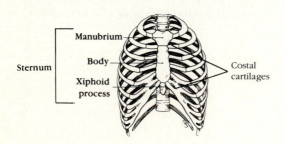

Sternum { Manubrium, Body, Xiphoid process }
Costal cartilages

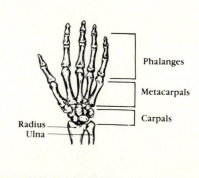

Superior border
Coracoid process
Spine
Glenoid fossa
Axillary border
Inferior border

18. **Figure 5-9:** Humerus.   19. **Figure 5-10:**   20. **Figure 5-11:**

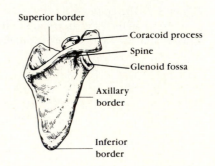

Greater tubercule
Lesser tubercule
Head
Shaft
Deltoid tuberosity
Capitulum
Trochlea

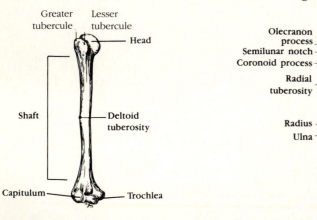

Olecranon process
Semilunar notch
Coronoid process
Radial tuberosity
Radius
Ulna

Phalanges
Metacarpals
Carpals
Radius
Ulna

21. Pectoral: A, C, D. Pelvic: B, E, F.
22. **1:** G or deltoid tuberosity.   **2:** I or humerus.   **3:** D or clavicle.   **4:** P or scapula.   **5:** O or radius.   **6:** T or ulna.   **7:** A or acromion process.   **8:** P or scapula.   **9:** D or clavicle.   **10:** H or glenoid fossa.   **11:** E or coracoid process.   **12:** D or clavicle.   **13:** S or trochlea.   **14:** T or ulna.   **15:** B or capitulum.   **16:** F or coronoid fossa.   **17:** T or ulna.   **18:** P or scapula.   **19:** Q or sternum.   **20:** C or carpals.   **21:** M or phalanges.   **22:** J or metacarpals.

23.   **1:** I or ilium, K or ischium, S or pubis.   **2:** J or ischial tuberosity.   **3:** R or pubic symphysis.   **4:** H or iliac crest.   **5:** A or acetabulum.   **6:** T or sacroiliac joint.   **7:** C or femur.   **8:** D or fibula.   **9:** W or tibia.   **10:** C or femur, W or tibia.   **11:** X or tibial tuberosity.   **12:** Q or patella.   **13:** W or tibia.   **14:** N or medial malleolus.   **15:** L or lateral malleolus.   **16:** B or calcaneus.   **17:** V or tarsals.   **18:** O or metatarsals.   **19:** P or obturator foramen.   **20:** G or greater and lesser trochanters, E or gluteal tuberosity.   **21:** U or talus.

24.   **1:** Female inlet is larger and more circular.   **2:** Female sacrum is less curved; pubic arch is rounder.   **3:** Female ischial spines are shorter; pelvis is shallower/lighter.

Figure 5-12:   25.   **Figure 5-13:**

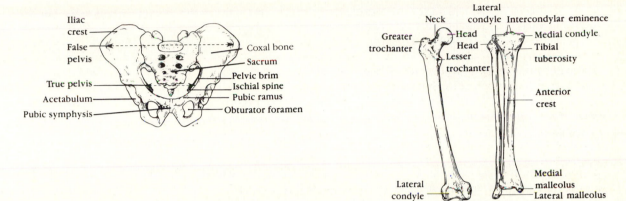

26.   **1:** Pelvic.   **2:** Phalanges.   **3:** *T.*   **4:** Acetabulum.   **5:** Sciatic.   **6:** *T.*   **7:** Hip bones.   **8:** *T.*

27.   **1:** Ulna.   **2:** Atlas.   **3:** Yellow marrow.   **4:** Styloid process.   **5:** Pelvis.   **6:** Scapula.   **7:** Mandible.   **8:** Condyle.   **9:** Sphenoid.   **10:** Marrow cavity.

28.   **Figure 5-14:** Bones of the skull, vertebral column, and body thorax are parts of the axial skeleton. All others belong to the appendicular skeleton.

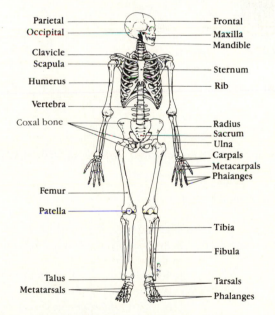

29.   **1:** F or simple fracture.   **2:** A or closed reduction.   **3:** D or greenstick fracture.   **4:** B or compression fracture.   **5:** C or compound fracture.   **6:** E or open reduction.

30.   **1:** *T.*   **2:** *T.*   **3:** Phagocytes (macrophages).   **4:** *T.*   **5:** Periosteum.   **6:** *T.*   **7:** Spongy.

31.   **1:** A or amphiarthrosis.   **2:** C or synarthrosis.   **3:** A or amphiarthrosis.   **4:** B or diarthrosis.   **5:** B or diarthrosis.   **6:** C or synarthrosis.   **7:** C or synarthrosis.   **8:** B or diarthrosis.   **9:** B or diarthrosis.   **10:** A or amphiarthrosis.   **11:** B or diarthrosis.   **12:** B or diarthrosis.   **13:** A or amphiarthrosis.   **14:** B or diarthrosis.   **15:** B or diarthrosis.

32.   **1:** Bursae.   **2:** Dislocation.   **3:** Mobility.   **4:** Synarthroses (sutures).

33.   **Figure 5-15:**

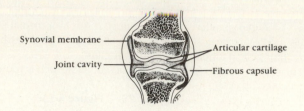

Synovial membrane — Articular cartilage
Joint cavity — Fibrous capsule

34.   **1:** *T.*   **2:** Osteoarthritis.   **3:** Acute.   **4:** Vascularized.   **5:** *T.*   **6:** Gouty arthritis or gout.   **7:** Rickets.   **8:** *T.*
35.   **1:** D or nervous.   **2:** F or urinary.   **3:** A or endocrine.   **4:** C or muscular.   **5:** A or endocrine.   **6:** B or integumentary.
36.   **1:** Fontanels.   **2:** Compressed.   **3:** Growth.   **4:** Sutures.   **5:** Thoracic.   **6:** Sacral.   **7:** Primary.   **8:** Cervical.
      9: Lumbar.
37.   **1:** Femur.   **2:** Spongy.   **3:** Stress (or tension).   **4:** RBCs (red blood cells).   **5:** Red marrow.   **6:** Nerve.
      7: Haversian.   **8:** Compact.   **9:** Canaliculi.   **10:** Lacunae (osteocytes).   **11:** Matrix.   **12:** Osteoclast.

Chapter 6

1.   **1:** A or cardiac, B or smooth.   **2:** A or cardiac, C or skeletal.   **3:** B or smooth.   **4:** C or skeletal.   **5:** A or cardiac.
     6: A or cardiac.   **7:** C or skeletal.   **8:** C or skeletal.   **9:** C or skeletal.
2.   **1:** G or perimysium.   **2:** B or epimysium.   **3:** I or sacromere.   **4:** D or myofiber.   **5:** A or endomysium.
     6: H or sarcolemma.   **7:** F or myofibril.   **8:** E or myofilament.   **9:** K or tendon.
     Figure 6-1: The endomysium is the connective
     tissue that surrounds each myofiber.

3.   **Figure 6-2:**

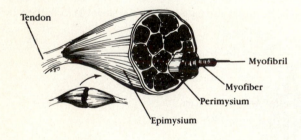

Tendon
Myofibril
Myofiber
Perimysium
Epimysium

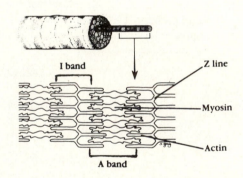

I band
Z line
Myosin
Actin
A band

4.   **1:** Neuromuscular.   **2:** Motor unit.   **3:** Axonal terminals.   **4:** Synaptic cleft.   **5:** Acetylcholine.   **6:** Nerve impulse
     (or action potential).   **7:** Depolarization.
5.   **1:** 1.   **2:** 4.   **3:** 7.   **4:** 2.   **5:** 5.   **6:** 3.   **7:** 6.
6.   **1:** F.   **2:** E.   **3:** C.   **4:** B.   **5:** H.   **6:** G.   **7:** I.
7.   Check #1, #3, #4, and #7.
8.   Your rate of respiration (breathing) is much faster and you breathe more deeply.
9.   **1:** G or tetanus.   **2:** B or isotonic contraction.   **3:** I or many motor units.   **4:** C or muscle cell.   **5:** J or repolarization.
     6: H or few motor units.   **7:** A or fatigue.   **8:** E or isometric contraction.
10. **1:** Insertion.   **2:** Origin.   **3:** Circumduction.   **4:** Adduct.   **5:** Flexion.   **6:** Extension.   **7:** Extension.   **8:** Flexed.
     9: Flexion.   **10:** Rotation.   **11:** Circumduction.   **12:** Rotation.   **13:** Plantar flexion.   **14:** Pronation.
     15: Abduction.   **16:** Dorsiflexion.
11. **1:** C or prime mover.   **2:** B or fixator.   **3:** D or synergist.   **4:** D or synergist.   **5:** A or antagonist.   **6:** B or fixator.
12. **1:** E, G.   **2:** A, G.   **3:** D, E.   **4:** E, F.   **5:** A, E.   **6:** B.   **7:** E, F.   **8:** E, F.

13. **Figure 6-3:**
Zygomatic;
Buccinator;
Orbicularis oculi;
Frontalis;
Orbicularis oris;
Masseter;
Temporalis;
Sternocleidomastoid.

14. **Figure 6-4:**
Rectus abdominis;
Pectoralis major;
Deltoid;
External oblique;
Sternocleidomastoid;
Internal oblique;
Transversus abdominis;
External intercostals.

15. **Figure 6-5:**
Trapezius;
Latissimus dorsi;
Deltoid.

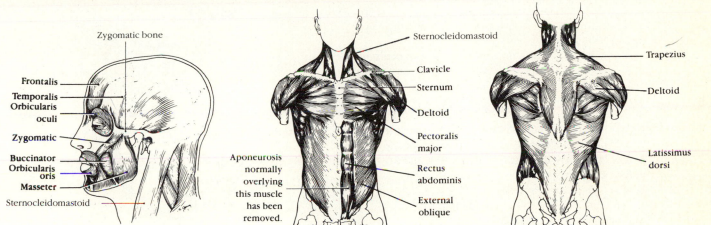

16. **Figure 6-6:** Iliopsoas; Gluteus maximus;
Gastrocnemius; Tibialis anterior;
Adductor group; Quadriceps;
Hamstrings; Gluteus medius; peroneus muscles.

17. **Figure 6-7:** Flexor carpi ulnaris;
Extensor digitorum;
Flexor digitorum superficialis;
Biceps brachii;
Triceps brachii;
Deltoid.

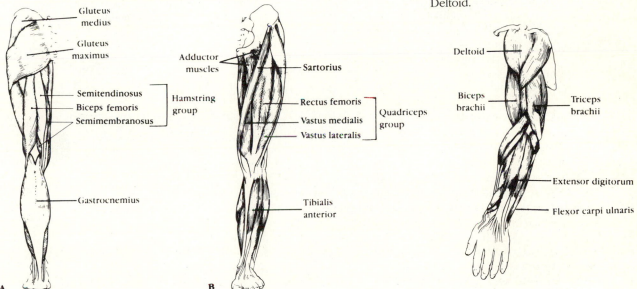

18. **1:** Deltoid. **2:** Gluteus maximus. **3:** Gluteus medius. **4:** Quadriceps. **5:** Achilles. **6:** Proximal. **7:** Forearm.
8: Anterior. **9:** Posteriorly. **10:** Thigh acting on leg.

19. **1:** 4. **2:** 5. **3:** 17. **4:** 16. **5:** 7. **6:** 6. **7:** 20. **8:** 14. **9:** 19. **10:** 12. **11:** 11. **12:** 10. **13:** 22. **14:** 1.
15: 2. **16:** 3. **17:** 15. **18:** 21. **19:** 13. **20:** 9. **21:** 18. **22:** 8.

20. **1:** 1. **2:** 2. **3:** 5. **4:** 9. **5:** 7. **6:** 4. **7:** 6. **8:** 3. **9:** 8. **10:** 10. **11:** 11.

21. **1:** Quickening. **2:** Muscular dystrophy. **3:** Proximal-distal. **4:** Cephalo-caudal. **5:** Gross. **6:** Fine.
7: Exercised. **8:** Atrophy. **9:** Myasthenia gravis. **10:** Weight. **11:** Size. **12:** Connective (scar).

22.   **1:** Endomysium.   **2:** Motor unit.   **3:** Myoneural (or neuromuscular).   **4:** Acetylcholine.   **5:** Sodium.
       6: Action potential.   **7:** Calcium   **8 and 9, in any order:** Actin; Myosin.   **10:** Calcium.

Chapter 7

1.   **1:** It monitors all information about changes occurring both inside and outside the body.   **2:** It processes and interprets the information received and integrates it in order to make decisions.   **3:** It commands responses by activating muscles, glands, and other parts of the nervous system.

2.   **1:** B or CNS.   **2:** D or somatic nervous system.   **3:** C or PNS.
       4: A or autonomic nervous system.   **5:** B or CNS.
       6: C or PNS.

3.   **1:** B or neuroglia.   **2:** A or neurons.   **3:** A or neurons.
       4: A or neurons.   **5:** B or neuroglia.

4.   **1:** B or axonal terminal.   **2:** C or dendrite.
       3: D or myelin sheath.   **4:** E or cell body.
       5: A or axon.

5.   **1:** A or bare nerve endings, G or Ruffini's corpuscle.
       2: A or bare nerve endings, F or Pacinian corpuscle.
       3: A or bare nerve endings (perhaps also B and E).
       4: B or golgi tendon organ, E or muscle spindle.
       5: D or Meissner's corpuscle.

6.   **1:** Stimulus.   **2:** Receptor.   **3:** Afferent neuron.
       4: Efferent Neuron.   **5:** Effector organ.

7.   **1:** C or cutaneous sense organs.
       2: L or Schwann cells.
       3: M or synapse.   **4:** O or tract.   **5:** B or association neuron.   **6:** I or nodes of Ranvier.   **7:** E or ganglion.
       8: D or efferent neuron.   **9:** K or proprioceptors.   **10:** N or stimuli.   **11:** A or afferent neuron.   **12:** G or neurotransmitters.

9.   **1:** E or refractory period.   **2:** B or depolarization.   **3:** C or polarized.   **4:** F or repolarization.   **5:** A or action potential.   **6:** D or potassium ions.   **7:** H or sodium-potassium pump.

10.   **1:** A or somatic reflex(es).   **2:** B or autonomic reflex(es).
       3: A or somatic reflex(es).   **4:** B or autonomic reflex(es).
       5: A or somatic reflex(es).   **6:** B or autonomic reflex(es).
       7: B or autonomic reflex(es).

11.   **1:** Pin-prick pain.   **2:** Skeletal muscle.
       3: Two (third with muscle).

8.   **Figure 7-1:**

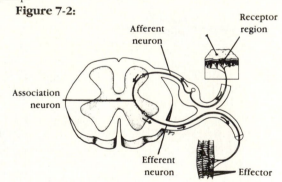

Cell body
Nucleus
Dendrites
Axon
Myelin sheath
Neurilemma
Axonal terminals
Node of Ranvier
Schwann cell nucleus

Figure 7-2:

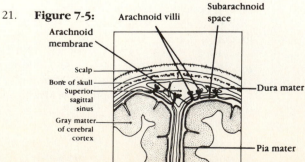

Afferent neuron
Receptor region
Association neuron
Efferent neuron
Effector

12.   **1:** Neurons.   **2:** K^+ enters the cell.
       3: Dendrite.   **4:** Unmyelinated.   **5:** Voluntary act.   **6:** Microglia.   **7:** Stretch.   **8:** High Na^+.

13.   **1:** Cerebral hemispheres.   **2:** Brain stem.   **3:** Cerebellum.   **4:** Ventricles.   **5:** Cerebrospinal fluid.

14.   Circle: Cerebral hemispheres; Cerebellum.

15.   **1:** Gyrus.   **2:** Surface area.   **3:** Neuron cell bodies.   **4:** Myelinated fibers.   **5:** Basal nuclei.

16.   **Figure 7-3, 1:** D.   **2:** L.   **3:** F.   **4:** C.   **5:** K.   **6:** B.   **7:** E.   **8:** A.   **9:** I.   **10:** H.   **11:** J,   **12:** G.   Areas B and C should be striped.

17.   **Figure 7-4, 1:** J.   **2:** L.   **3:** O.   **4:** M.   **5:** B.   **6:** A.   **7:** K.   **8:** G.   **9:** I.   **10:** E.   **11:** N.   **12:** F.   **13:** H.
       14: C.   Structures #4, #9, and #13 should be blue. Structures #2, #7, the cavity enclosed by #14 and #8, and the entire gray area around the brain should be colored yellow.

18.   **1:** Hypothalamus.   **2:** Pons.   **3:** Cerebellum.
       4: Thalamus.   **5:** Medulla oblongata.   **6:** Corpus callosum.
       7: Cerebral aqueduct.   **8:** Thalamus.   **9:** Choroid plexi.
       10: Cerebral peduncle.   **11:** Hypothalamus.

19.   **1:** Postcentral.   **2:** Temporal.   **3:** Frontal.   **4:** Broca's.
       5: Left.   **6:** *T.*   **7:** Reticular system.   **8:** *T.*   **9:** Alert.

20.   **1:** Dura mater.   **2:** Pia mater.   **3:** Arachnoid villi.
       4: Arachnoid.   **5:** Dura mater.

21.   **Figure 7-5:**

Arachnoid villi
Subarachnoid space
Arachnoid membrane
Scalp
Bone of skull
Superior sagittal sinus
Gray matter of cerebral cortex
Dura mater
Pia mater

22. **1:** Choroid plexi.   **2:** Ventricles.   **3:** Cerebral aqueduct.   **4:** Central canal.   **5:** Subarachnoid space.
    6: Fourth ventricle.   **7:** Hydrocephalus.
23. **1:** Cerebellum.   **2:** Basal nuclei.   **3:** Meningitis.   **4:** III (oculomotor).   **5:** Somatosensory cortex.   **6:** Broca's area.
    7: Electroencephalogram.
24. **1:** Foramen magnum.   **2:** Lumbar.   **3:** Lumbar tap, or puncture.   **4:** 31.   **5:** 8.   **6:** 12.   **7:** 5.   **8:** 5.
    9: Cauda equina.
25. **1:** D or association neurons.   **2:** B or efferent.   **3:** A or afferent.   **4:** B or efferent.   **5:** A or afferent.
    6: C or both afferent and efferent.   **7:** C or both afferent and efferent.
26. **Figure 7-6:**

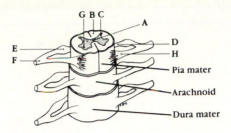

27. **1:** B or loss of sensory function.   **2:** A or loss of motor function.   **3:** C or loss of both sensory and motor function.

28. **Figure 7-7:**

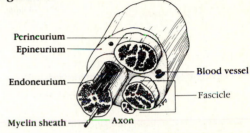

29. **1:** Nerve (or fascicle).   **2:** Mixed.   **3:** Afferent.
30. **Figure 7-8:**

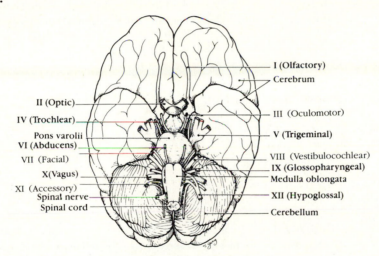

31. **1:** XI-Accessory.   **2:** I-Olfactory.   **3:** III-Oculomotor.   **4:** X-Vagus.   **5:** VII-Facial.   **6:** V-Trigeminal.
    7: VIII-Vestibulocochlear.   **8:** VII-Facial.   **9:** III, IV, VI.   **10:** V-Trigeminal.   **11:** II-Optic.   **12:** I, II, VIII.
32. **1:** Plexuses.   **2:** Limbs.   **3:** Thorax.   **4:** Posterior body trunk.
33. **1:** Cervical plexus.   **2:** Phrenic nerve.   **3:** Sciatic nerve.   **4:** Peroneal and tibial nerves.   **5:** Median nerve.
    6: Musculocutaneous nerve.   **7:** Lumbar plexus.   **8:** Femoral nerve.   **9:** Ulnar nerve.
34. Check sympathetic for 1, 4, 6, 8, and 10. Check parasympathetic for 2, 3, 5, 7, 9, and 11.
35. **1:** Increased respiratory rate.   **2:** Increased heart rate and blood pressure.   **3:** Increased availability of blood glucose.
    4: Pupils dilate; increased blood flow to heart, brain, and skeletal muscles.
36. Parasympathetic.
37. **1:** Hypothalamus.   **2:** Oxygen.   **3:** Cephalocaudal.   **4:** Gross.   **5:** Blood pressure.   **6:** Decreased oxygen (blood)
    to brain.   **7:** Senility.   **8:** Stroke (CVA).
38. **1:** Cerebellum.   **2:** Medulla.   **3:** Hypothalamus.   **4:** Memories.   **5:** Temporal.   **6:** Broca's area.   **7:** Reasoning.
    8: Frontal.   **9:** Vagus (X).   **10:** Dura mater.   **11:** Subarachnoid space.   **12:** Fourth.

Chapter 8

1. **1:** Extrinsic, or external.   **2:** Eyelids.   **3:** Ciliary glands.   **4:** Sty.
2. **1:** 2.   **2:** 4.   **3:** 3.   **4:** 1.

3. **Figure 8-1:**
   1: Superior rectus turns eye superiorly.   **2:** Inferior rectus turns eye inferiorly.   **3:** Superior oblique turns eye inferiorly and laterally.   **4:** Lateral rectus turns eye laterally.   **5:** Medial rectus turns eye medially.   **6:** Inferior oblique turns eye superiorly and laterally.

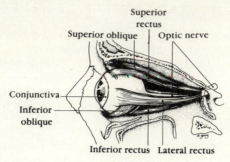

4. **1:** Conjunctiva secretes mucus.   **2:** Lacrimal glands secrete salt water and lysozyme.   **3:** Meibomian and ciliary glands secrete oil. Circle the lacrimal gland secretion.

5. **1:** L or refraction.   **2:** A or accommodation.   **3:** F or emmetropia.   **4:** H or hyperopia.   **5:** K or photopupillary reflex.   **6:** D or cataract.   **7:** I or myopia.   **8:** C or astigmatism.   **9:** G or glaucoma.   **10:** E or convergence.   **11:** B or accommodation pupillary reflex.   **12:** J or night blindness.

6. **1:** Autonomic nervous system.

7. **1:** Convex.   **2:** Real.   **3:** Behind.   **4:** Convex (converging).   **5:** In front of.   **6:** Concave (diverging).

8. **1:** L or suspensory ligaments.   **2:** A or aqueous humor.   **3:** K or sclera.   **4:** I or optic disk.   **5:** D or ciliary body.   **6:** C or choroid coat.   **7:** B or canal of Schlemm.   **8:** J or retina.   **9:** M or vitreous humor.   **10:** C or choroid coat.   **11:** D or ciliary body.   **12:** G or iris.   **13:** F or fovea centralis.   **14-17:** A or aqueous humor, E or cornea, H or lens, and M or vitreous humor.   **18:** E or cornea.   **19:** K or sclera.

9. **Figure 8-2:**

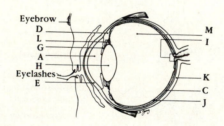

10. **1:** In distant vision the ciliary muscle is relaxed, the lens convexity is decreased, and the degree of light refraction is decreased.   **2:** In close vision the ciliary muscle is contracted, the lens convexity is increased, and the degree of light refraction is increased.

11. Retina → Optic nerve → Optic chiasma → Optic tract → Synapse in thalamus → Optic radiation → Optic cortex.

12. **1:** Three.   **2-4:** Blue, green, and red.   **5:** At the same time.   **6:** Total color blindness.   **7:** Males.   **8:** Rods.

13. **1:** Vitreous humor.   **2:** Superior rectus.   **3:** Far vision.   **4:** Mechanoreceptors.   **5:** Iris.   **6:** Conjunctiva.

14. **1-3:** E or external auditory canal, I or pinna, and M or tympanic membrane.   **4-6:** B or cochlea, K or semicircular canals, and N or vestibule.   **7-9:** A or anvil, F or hammer, and L or stirrup.   **10 and 11:** K or semicircular canals and N or vestibule.   **12:** D or eustachian tube.   **13:** M or tympanic membrane.   **14:** B or cochlea.   **15:** D or eustachian tube.   **16 and 17:** K or semicircular canals and N or vestibule.   **18:** G or oval window.   **19:** C or endolymph.   **20:** H or perilymph.

15. **Figure 8-3:** I, E, and M are yellow; A, F, and L are red; B is blue, and K (continuing to J) is green.

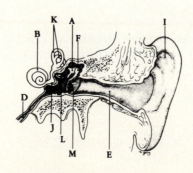

16. Eardrum →Hammer →Anvil →Stirrup →Oval window →Perilymph →Membrane →Endolymph →Hair cells.
17. **Figure 8-4:**

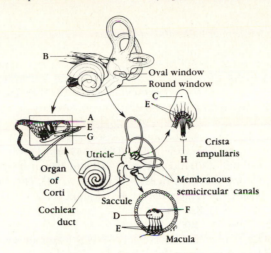

18. **1:** C or dynamic. **2:** I or semicircular canals. **3:** A or angular/rotary. **4:** D or endolymph. **5:** B or cupula. **6:** J or static. **7:** H or saccule. **8:** K or utricle. **9:** E or gravity. **10:** L or vision. **11:** G or proprioception.
19. **1:** C. **2:** S. **3:** S **4:** C. **5:** C, S. **6:** C. **7:** S.
20. Nausea, dizziness, and balance problems.
21. **1:** Pinna. **2:** Tectorial membrane. **3:** Sound waves. **4:** Eustachian tube. **5:** Optic nerve. **6:** Retina.
22. **1-3:** (in any order): VII-Facial, IX-Glossopharyngeal; X-Vagus. **4:** I-Olfactory. **5:** Mucosa. **6:** Sniffing. **7:** Taste buds. **8:** Fungiform. **9:** Circumvallate. **10-13:** (in any order): Sweet; Salty; Bitter; Sour. **14:** bitter. **15:** Smell. **16:** Dry. **17:** Memories.
23. **Figure 8-5:** The predominant site of sweet receptors is on the tip of the tongue. As you progress down the sides of the tongue, salt receptors become more numerous, followed by sour receptors. Bitter receptors are located at the back of the tongue.

24. **Figure 8-6:**

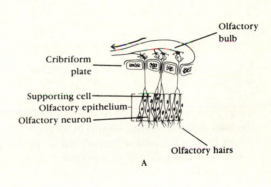

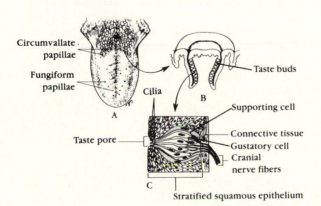

25. **1:** Musky. **2:** Epithelial cell. **3:** Neuron. **4:** Olfactory nerve. **5:** Four receptor types. **6:** Metal ions. **7:** Photoreceptors.
26. **1:** Nervous system. **2:** Measles (rubella). **3:** Blindness. **4:** Vision. **5:** Hyperopic. **6:** Elastic. **7:** Presbyopia. **8:** Cataract. **9:** Presbycusis.
27. **1:** Bony labyrinth. **2:** Perilymph. **3:** Saccule. **4:** Utricle. **5:** Gel. **6:** Otoliths. **7:** Macula. **8:** Static. **9:** Cochlear duct. **10:** Organ of Corti. **11:** Hearing. **12:** Cochlear division of cranial nerve VIII. **13:** Semicircular canals. **14:** Crista ampullaris. **15:** Dynamic.

Chapter 9

1. **1:** Steroid.   **2:** Receptor.   **3:** Messenger RNAs and/or proteins.   **4:** First.   **5:** Adenylate cyclase.   **6:** ATP. **7:** Second.

2. **1:** P or slower.   **2:** L or nervous system.   **3:** F or hormones.   **4:** K or nerve impulses.   **5:** D or circulatory system. **6:** A or all body cells.   **7:** N or receptors.   **8:** T or target cell(s).   **9:** B or altering activity.   **10:** R or stimulating new or unusual activities.   **11:** Q or steroid or amino acid-based.   **12:** M or neural.   **13:** E or hormonal. **14:** G or humoral.   **15:** J or negative feedback.   **16:** C or anterior pituitary.   **17:** O or releasing hormones. **18:** H or hypothalamus.

3. **1:** Thyroxine.   **2:** Thymosin.   **3:** PTH.   **4:** Cortisone (glucocorticoids).   **5:** Epinephrine.   **6:** Insulin.   **7-10:** (in any order): TSH; FSH; LH; ACTH.   **11:** Glucagon.   **12:** ADH.   **13 and 14:** (in any order): FSH; LH.   **15 and 16:** (in any order): Estrogen; Progesterone.   **17:** Aldosterone.   **18 and 19:** (in any order): Prolactin and oxytocin.

4. **Figure 9-1, A:** Pineal.   **B:** Posterior pituitary.   **C:** Anterior pituitary.   **D:** Thyroid.   **E:** Thymus.   **F:** Adrenal. **G:** Pancreas.   **H:** Ovary.   **I:** Testis.   **J:** Parathyroids.   **K:** Placenta.

5. **1:** C.   **2:** B.   **3:** F.   **4:** F.   **5:** F.   **6:** H, K.   **7:** C.   **8:** G.   **9:** G.   **10:** C.   **11:** A.   **12:** B.   **13:** H, K.   **14:** C. **15:** J.   **16:** C.   **17:** I.   **18:** E.   **19:** D.   **20:** D.   **21:** C.

6. **1:** Kidneys; Causes retention of Ca^{2+} and enhances excretion of PO_4^{3-}. Promotes activation of vitamin D.   **2:** Intestine: Causes increased absorption of Ca^{2+} from foodstuffs.   **3:** Bones: Causes enhanced release of Ca^{2+} from bone matrix.

7. **1:** Estrogen/testosterone.   **2:** PTH.   **3:** ADH.   **4:** Thyroxine.   **5:** Thyroxine.   **6:** Insulin.   **7:** Growth hormone. **8:** Estrogen/progesterone.   **9:** Thyroxine.

8. **1:** Growth hormone.   **2:** Thyroxine.   **3:** PTH.   **4:** Glucocorticoids.   **5:** Growth hormone.   **6:** Androgens.

9. **1:** Polyuria—high sugar content in kidney filtrate causes large amount of water to be lost in the urine.   **2:** Polydipsia— thirst due to large volumes of urine excreted.   **3:** Polyphagia—hunger because blood sugar cannot be used as a body fuel even though levels are high.

10. **1:** Neoplasm.   **2:** Hypersecretion.   **3:** Iodine.   **4:** Estrogen.   **5:** Menopause.   **6:** Bear children.   **7:** Insulin.

11. **1:** Insulin.   **2:** Pancreas.   **3:** Posterior pituitary, or hypothalamus.   **4:** ADH.   **5:** Parathyroid.   **6:** Calcium. **7:** Adrenal medulla.   **8:** Epinephrine.   **9:** Thyroxine.

Chapter 10

1. **1:** Connective tissue.   **2:** Formed elements.   **3:** Plasma.   **4:** Clotting.   **5:** Erythrocytes.   **6:** Hematocrit. **7:** Plasma.   **8:** Leukocytes.   **9:** Platelets.   **10:** One.   **11:** Oxygen.

2. **1:** F or neutrophil.   **2-4** (in any order): C or eosinophil, D or basophil, F or neutrophil.   **5:** A or red blood cell. **6 and 7** (in any order): E or monocyte, F or neutrophil.   **8 and 9** (in any order): E or monocyte, G or lymphocyte. **10:** B or megakaryocyte.   **11:** H or formed elements.   **12:** C or eosinophil.   **13:** D or basophil.   **14:** G or lymphocyte.   **15:** A or red blood cell.   **16:** I or plasma.   **17:** E or monocyte.   **18:** D or basophil.   **19-23** (in any order): C or eosinophil, D or basophil, E or monocyte, F or neutrophil, G or lymphocyte.

3.   **Figure 10-1:** 100–120 days.

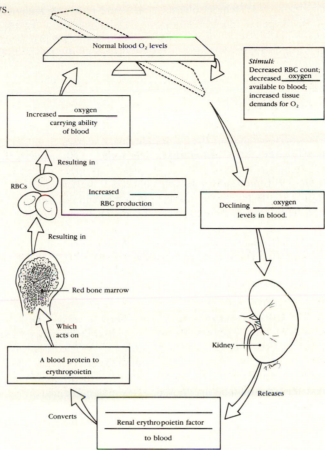

Normal blood O₂ levels

Stimuli:
Decreased RBC count;
decreased ___oxygen___
available to blood;
increased tissue
demands for O₂

Increased ___oxygen___
carrying ability
of blood

Resulting in

RBCs

Increased _____
RBC production

Declining ___oxygen___
levels in blood.

Resulting in

Red bone marrow

Which
acts on

Kidney

A blood protein to
erythropoietin

Releases

Converts

Renal erythropoietin factor
to blood

4.   **Figure 10-2:** Neutrophil.   **Figure 10-3:** Monocyte.   **Figure 10-4:** Eosinophil.   **Figure 10-5:** Lymphocyte.

5.   **1:** Diapedesis.   **2:** Differential.   **3:** Kidneys.   **4:** 7.35.   **5:** 5.5.   **6:** T.   **7:** O.   **8:** 4.5-5.5.   **9:** Hematocrit.
     10: Less.   **11:** Heparin.   **12:** Monocytes.   **13:** Lymphocytes.   **14:** 2-6.   **15:** Polycythemia.   **16:** T.

6.   **1:** A or break.   **2:** E or platelets.   **3:** G or serotonin.   **4:** I or thromboplastin.   **5:** F or prothrombin.
     6: H or thrombin.   **7:** D or fibrinogen.   **8:** C or fibrin.   **9:** B or erythrocytes.

7.   **1:** Erythrocytes.   **2:** Monocytes.   **3:** Antibodies.   **4:** Lymphocyte.   **5:** Thrombocytes.   **6:** Aneurysm.
     7: Increased hemoglobin.   **8:** Hemoglobin.   **9:** Lymphocyte.

8.

Blood type	Agglutinogens or antigens	Agglutinins or antibodies in plasma	Can donate blood to type	Can receive blood from type
1: Type A	A	anti-B	A, AB	A, O
2: Type B	B	anti-A	B, AB	B, O
3: Type AB	A, B	none	AB	A, B, AB, O
4: Type O	none	anti-A, anti-B	A, B, AB, O	O

9.   No; A+

10. Plasma antibodies attach to and lyse red blood cells different from your own.

11. **1:** Hemolytic disease of the newborn.   **2:** RBCs have been destroyed by the mother's antibodies; therefore, the baby's
     blood is carrying insufficient oxygen.   **3:** She must have received mismatched (Rh+) blood previously in a
     transfusion.   **4:** Give the mother RhoGAM to prevent her from becoming sensitized to the Rh+ antigen.

12. **1:** F.   **2:** Jaundiced.   **3:** Sickle cell.   **4:** Hemophilia.   **5:** Iron.   **6:** Pernicious.   **7:** B₁₂.   **8:** Thrombi.
     9: Leukemia.

13.   **1:** Hemopoiesis.   **2:** Hemostasis.   **3:** Hemocytoblasts.   **4:** Neutrophil.   **5:** Phagocyte.   **6:** REF.   **7:** Red blood cells.   **8:** Hemoglobin.   **9:** Oxygen.   **10:** Lymphocytes.   **11:** Antibodies.   **12-15:** (in any order): Basophils; Eosinophils; Monocytes; Thrombocytes (platelets).   **16:** Endothelium.   **17:** Platelets.   **18:** Serotonin.   **19:** Fibrin.   **20:** Clot.   **21:** Thromboplastin.   **22:** Prothrombin.   **23:** Thrombin.   **24:** Fibrinogen.   **25:** Embolus.

Chapter 11

1.   **1:** Right ventricle.   **2:** Pulmonary semilunar valve.   **3:** Pulmonary arteries.   **4:** Lungs.   **5:** R and L pulmonary veins.   **6:** Left atrium.   **7:** Mitral or bicuspid.   **8:** Left ventricle.   **9:** Aortic.   **10:** Capillary beds.   **11:** Superior vena cava.   **12:** Inferior vena cava.

2.   **Figure 11-1, 1:** Right atrium.   **2:** Left atrium.   **3:** Right ventricle.   **4:** Left ventricle.   **5:** Superior vena cava.   **6:** Inferior vena cava.   **7:** Aorta.   **8:** Pulmonary trunk.   **9:** Left pulmonary artery.   **10:** Right pulmonary artery.   **11:** Right pulmonary veins.   **12:** Left pulmonary veins.   **13:** Coronary circulation.   **14:** Apex of heart.   **15:** Ligamentum arteriosum.

3.   **1:** Thorax.   **2:** Diaphragm.   **3:** Second.   **4:** Right atrium.   **5:** Right ventricle.   **6:** Left atrium.   **7:** Aorta.   **8:** Right atrium.   **9:** Left ventricle.   **10:** Atria.   **11:** Ventricles.   **12:** Endocardium.   **13:** Epicardium.   **14:** Friction.   **15:** Cardiac muscle.

4.   **1:** Systole.   **2:** Diastole.   **3:** Lub-dup.   **4:** Atrioventricular.   **5:** Semilunar.   **6:** Ventricles.   **7:** Atria.   **8:** Atria.   **9:** Ventricles.   **10:** Murmurs.

5.   **A and B:** (in any order): 6, 7.   **C and D:** (in any order): 8, 9.   **E:** 9.   **F:** 8   **G:** 1.   **H:** 2.   **1:** SA node.   **2:** AV node.   **3:** AV bundle or bundle of His.   **4:** Bundle branches.   **5:** Purkinje fibers.   **6:** Pulmonary valve.   **7:** Aortic valve.   **8:** Mitral (bicuspid) valve.   **9:** Tricuspid valve.
     Figure 11-2: Red arrows should be drawn from the left atrium to the left ventricle and out the aorta. Blue arrows should be drawn from the superior and inferior vena cavae into the right atrium then into the right ventricle and out the pulmonary trunk. Green arrows should be drawn from #1 to #5 in numerical order.

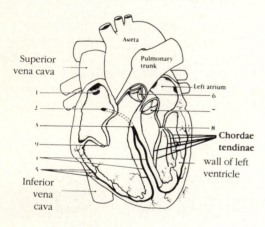

6.   **1:** C or electrocardiogram.   **2:** F or P wave.   **3:** H or T wave.   **4:** G or QRS wave.   **5:** B or bradycardia.   **6:** D or fibrillation.   **7:** I or tachycardia.   **8:** E or heart block.   **9:** A or angina pectoris.

7.   **Figure 11-3:**

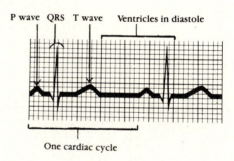

8.   **1:** Cardiac output.   **2:** Heart rate.   **3:** Stroke volume.   **4:** About 75 beats per minute.   **5:** 70 ml per beat.   **6:** 5,250 ml per minute.   **7:** Minute.   **8:** Stroke volume.   **9:** Stretch.   **10:** Blood.

9.   Check 1, 2, 4, 5, 6, 9, and 10.

10. **1:** Fetal.   **2:** Rate.   **3:** Left.   **4:** *T.*   **5:** *T.*

11. **1:** Left side of heart.   **2:** P wave.   **3:** AV valves opened.   **4:** Aortic semilunar valve.   **5:** Tricuspid valve.   **6:** Coronary sinus.   **7:** Vagus nerves.   **8:** Heart block.

12. **1:** Lumen.   **2:** Vasoconstriction.   **3:** Vasodilation.   **4:** Veins.   **5:** Arteries.   **6:** Arterioles.   **7:** Venules.

13. Arteries are high-pressure vessels. Veins are low-pressure vessels. Blood flows from high to low pressure. The venous valves help to prevent the backflow of blood that might otherwise occur in those low-pressure vessels.

14. Skeletal muscle activity and breathing (respiratory pump).

15. **1:** A or tunica intima.   **2:** B or tunica media.   **3:** A or tunica intima.   **4:** A or tunica intima.   **5:** C or tunica externa.   **6:** B or tunica media.   **7:** C or tunica externa. **Figure 11-4, A:** Artery; thick media; small, round lumen. **B:** Vein; thin media; elongated, relatively collapsed lumen. **C:** Capillary; single layer of endothelium. The tunica intima is the innermost vessel layer; the tunica externa is the outermost layer; and the tunica media is the thick middle layer.

16. **Figure 11-5:**

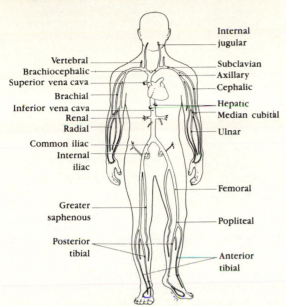

Figure 11-6:

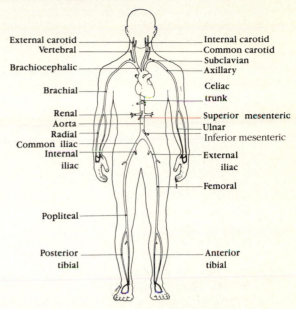

17. **1 and 2:** S or radial and X or ulnar. **3:** U or subclavian. **4:** E or cardiac. **5:** T or renal. **6:** Q or internal jugular. **7:** D or brachiocephalic. **8 and 9:** A or anterior tibial and R or posterior tibial. **10:** M or hepatic portal. **11:** F or cephalic. **12:** J or gonadal. **13:** B or azygos. **14:** O or inferior vena cava. **15:** L or hepatic. **16-18:** I or gastric, N or inferior mesenteric, and V or superior mesenteric. **19:** K or greater saphenous. **20:** G or common iliac. **21:** H or femoral.

18. **1 and 2:** F or common carotid and W or subclavian. **3:** H or coronary. **4 and 5:** P or internal carotid and Y or vertebral. **6:** B or aorta. **7:** J or dorsalis pedis. **8:** I or deep femoral. **9:** S or phrenic. **10:** C or brachial. **11:** C or brachial. **12:** N or inferior mesenteric. **13:** Q or internal iliac. **14:** L or femoral. **15:** C or brachial. **16:** X or superior mesenteric. **17:** G or common iliac. **18:** E or celiac trunk. **19:** K or external carotid. **20-22:** (in any order): A or anterior tibial, R or peroneal, T or posterior tibial. **23:** U or radial.

19. **Figure 11-7:** 20. **Figure 11-8:**

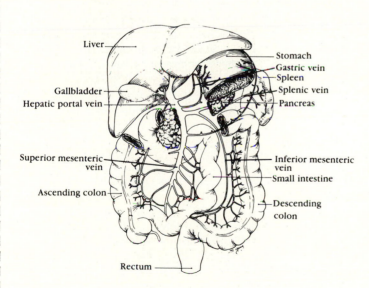

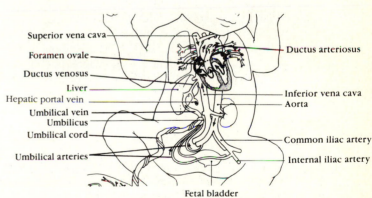

21. **Figure 11-9:** (1) Left internal carotid artery to left anterior cerebral artery. Then through anterior communicating branch to right anterior cerebral artery and (2) vertebral arteries to basilar artery to right posterior cerebral artery through the posterior communicating branch to right middle cerebral artery.

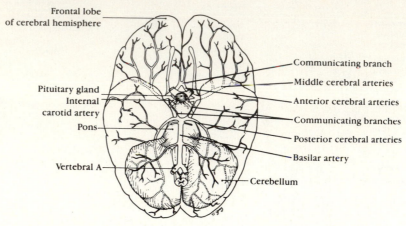

22. **1:** C or circle of Willis.   **2:** J or umbilical vein.   **3:** E or ductus venosus.   **4:** A or anterior cerebral artery, G or middle cerebral artery.   **5:** B or basilar artery.   **6:** D or ductus arteriosus.   **7:** F or foramen ovale.

23. The fetal lungs are not functioning in gas exchanges, and they are collapsed. The placenta makes the gas exchanges with the fetal blood.

24. **1:** Aorta.   **2:** Carotid artery.   **3:** Vasodilation.   **4:** High blood pressure.   **5:** Vasodilation.   **6:** Vein.

25. **1:** D.   **2:** I.   **3:** I.   **4:** I.   **5:** I.   **6:** D.   **7:** D.   **8:** I.   **9:** D.   **10:** D.   **11:** D.   **12:** I.   **13:** I.

26. **1:** Femoral artery.   **2:** Brachial artery.   **3:** Popliteal artery.   **4:** Facial artery.   **5:** Radial artery.

27. **1:** Increase.   **2:** Orthostatic.   **3:** Brain.   **4:** Stethoscope.   **5:** Low.   **6:** *T.*   **7:** Vasoconstricting.   **8:** Hypertension.

28. **1:** H or pulse.   **2:** B or blood pressure.   **3 and 4:** C or cardiac output and F or peripheral resistance.   **5:** D or constriction of arterioles.   **6:** J or systolic blood pressure.   **7:** E or diastolic blood pressure.   **8:** A or over arteries.   **9:** G or pressure points.   **10:** I or sounds of Korotkoff.

29. **1:** C or spleen.   **2:** A or lymph nodes.   **3:** D or thymus.   **4:** B or Peyer's patches, E or tonsils.   **5:** C or spleen.   **6:** B or Peyer's patches.   **7:** C or spleen.

30. **1:** Heart (pump).   **2:** Arteries.   **3:** Veins.   **4:** Valves.   **5:** Lymph.   **6:** Right lymphatic duct.   **7:** Thoracic duct.   **8:** Lymph nodes.   **9-11:** (in any order): Cervical; Inguinal; Axillary.   **12:** Protect.

31. **1:** Fourth.   **2:** Occluded.   **3:** Death.   **4:** Aerobic exercise.   **5:** Arteriosclerosis.   **6:** Varicose veins.   **7 and 8:** Feet and legs.   **9 and 10:** Hypertension; Coronary artery disease.   **11:** Heart attacks.   **12-15:** (in any order): Animal fat/cholesterol; Salt; Smoking; Lack of exercise.   **16:** Veins.   **17:** Thymus.

32. **1:** Left atrium.   **2:** Left ventricle.   **3:** Mitral.   **4:** Chordae tendinae.   **5:** Diastole.   **6:** Systole/contraction.   **7:** Aortic semilunar.   **8:** Aorta.   **9:** Superior mesenteric.   **10:** Endothelial.   **11:** Superior mesenteric vein.   **12:** Splenic.   **13:** Nutrients.   **14:** Phagocytic (van Kupffer).   **15:** Hepatic.   **16:** Inferior vena cava.   **17:** Right atrium.   **18:** Pulmonary.   **19:** Lungs.   **20:** Subclavian.

Chapter 12

1.   **1:** Tears and saliva.   **2:** Stomach and female reproductive tract.   **3:** Sebaceous glands; skin.   **4:** Digestive.

2.   **1:** A or acids, B or lysozyme, F or sebum.   **2:** C or mucosae, G or skin.   **3:** A or acids, B or lysozyme, D or mucus, E or protein-digesting enzymes, F or sebum.   **4:** D or mucus.   **5:** A-G.

3.   They propel mucus laden with trapped debris superiorly away from the lungs to the throat, where it can be spat out.

4.   Phagocytosis is ingestion and destruction of particulate material. The rougher the particle, the more easily it is ingested.

5.   **1:** Itching.   **2:** Natural killer cells.   **3:** Interferon.   **4:** Inflammation.   **5:** Anti-bacterial.

6.   **1:** Prevents the spread of damaging agents to nearby areas.   **2:** Disposes of cell debris and pathogens.   **3:** Sets the stage for repair (healing).

7.   **1:** F or increased blood flow.   **2:** E or histamine.   **3:** G or inflammatory chemicals.   **4:** A or chemotaxis.   **5:** C or edema.   **6:** H or macrophages.   **7:** B or diapedesis.   **8:** I or neutrophils.   **9:** D or fibrin mesh.

8.   **1:** Proteins.   **2:** Activated.   **3:** Membrane Attack Complex (MAC).   **4:** Hole or lesion.   **5:** Lysis.   **6:** Opsonization.

9.   Interferon is synthesized in response to viral infection of a cell. The cell produces and releases interferon proteins, which diffuse to nearby cells, where they prevent viruses from multiplying within those cells.

10. **1:** A or antigens.   **2:** E or humoral immunity.   **3:** D or cellular immunity.   **4 and 5:** B or B cells and I or T cells.   **6:** H or macrophages.   **7 and 8:** C or blood and F or lymph.   **9:** G or lymph nodes.

11. **Figure 12-1:**

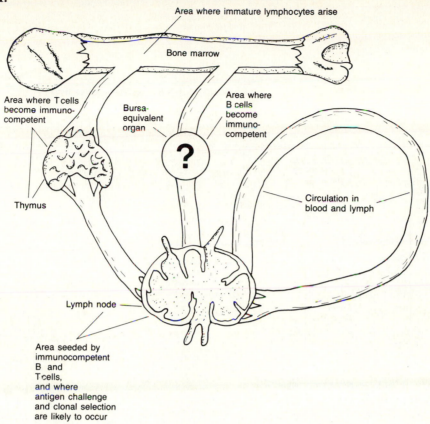

Area where immature lymphocytes arise

Bone marrow

Area where T cells become immuno-competent

Bursa-equivalent organ

Area where B cells become immuno-competent

?

Circulation in blood and lymph

Thymus

Lymph node

Area seeded by immunocompetent B and T cells, and where antigen challenge and clonal selection are likely to occur

1: The appearance of antigen-specific receptors on the membrane of the lymphocyte. **2:** Fetal life. **3:** Its genes. **4:** Binding to "its" antigen. **5:** Bone marrow. **6:** "Self."

12.

Characteristic	T cell	B cell
Originates in bone marrow from stem cells called hemocytoblasts	√	√
Progeny are plasma cells		√
Progeny include suppressors, helpers, and killers	√	
Progeny include memory cells	√	√
Is responsible for directly attacking foreign cells or viruses that enter the body	√	
Produces antibodies that are released to body fluids		√
Bears a cell-surface receptor capable of recognizing a specific antigen	√	√
Forms clones upon stimulation	√	√
Accounts for most of the lymphocytes in the circulation	√	

13. **1:** Immune system. **2:** Proteins. **3:** Haptens. **4:** Nonself.
14. **1:** A or helper T cell. **2:** A or helper T cell. **3:** C or suppressor T cell. **4:** B or killer T cell.
15. **1:** F or interferon. **2:** C or chemotaxis factors. **3:** B or antibodies. **4:** E or inflammation. **5:** G or lymphokine. **6:** D or complement. **7:** I or monokines.

16.   **Figure 12-2:**

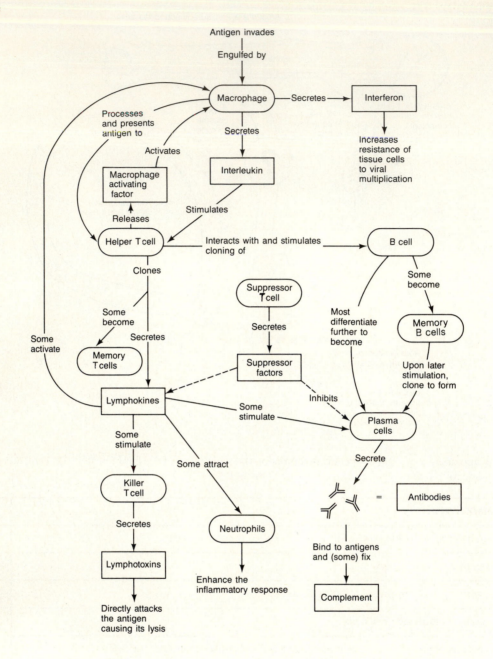

17.   **1:** Lymphokines.   **2:** Hapten.   **3:** Liver.
18.   **Figure 12-3:**
      1: The V portion.   **2:** The C portion.

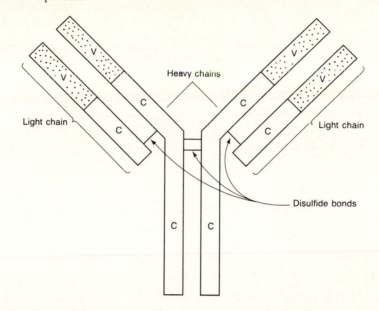

19.   **1:** B or IgD.   **2:** D or IgG.   **3:** E or IgM.   **4:** D or IgG, E or IgM.   **5:** E or IgM.   **6:** D or IgG.   **7:** C or IgE.
     8: A or IgA.
20.   **1:** Antigen.   **2:** Complement activation and lysis.   **3:** Neutralization.   **4:** Agglutination.   **5:** IgM.   **6:** Precipitation.
     7: Phagocytes.
21.   **1:** A.   **2:** P.   **3:** P.   **4:** P.   **5:** A.   **6:** A.
22.   **1:** P.   **2:** P.   **3:** S.   **4:** P.   **5:** S.
23.   **1:** C or immunodeficiency.   **2:** A or allergy.   **3:** A or allergy.   **4:** C or immunodeficiency.   **5:** B or autoimmune
     disease.   **6:** C or immunodeficiency.   **7:** B or autoimmune disease.   **8:** A or allergy.   **9:** A or allergy.
24.   **1:** Liver.   **2:** Lymphatic organs.   **3:** Thymus.   **4:** Thymosin.   **5:** Birth (or shortly thereafter).   **6:** Depressed.
     7: Nervous.   **8:** Declines.   **9-11** (in any order): Immunodeficiencies; Autoimmune diseases; Cancer.   **12:** IgA.
25.   **1:** Protein.   **2:** Lymph node.   **3:** B lymphocytes (B cells).   **4:** Plasma cell.   **5:** Antibodies.   **6:** Macrophage.
     7: Antigens.   **8:** Antigen presenters.   **9:** T.   **10:** Clone.   **11:** Immunologic memory.

Chapter 13

1.   **1:** R.   **2:** L.   **3:** R.
2.   **1:** External nares.   **2:** Nasal septum.   **3-5** (in any order): Warm; Moisten; Trap debris in.   **6:** Sinuses.   **7:** Speech.
     8: Pharynx.   **9:** Larynx.   **10:** Tonsils.   **11:** Cartilage.   **12:** Pressure.   **13:** Anteriorly.   **14:** Cricoid.
     15: Thyroid.   **16:** Vocal cords.   **17:** Speak.
3.   **1:** Mandibular.   **2:** Alveolus.   **3:** Larynx.   **4:** Peritonitis.   **5:** Nasopharynx.

4.   **Figure 13-1:**

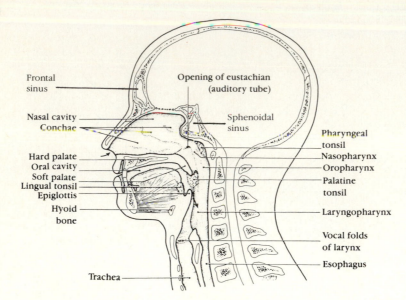

5.   **1:** B or bronchioles.   **2:** G or palate.   **3:** I or phrenic nerve.   **4:** E or esophagus.   **5:** D or epiglottis.
       6: K or trachea.   **7:** A or alveoli.   **8:** H or parietal pleura.   **9:** L or vagus nerve.   **10:** F or glottis.
       11: C or conchae.
6.   **1:** Elastic connective.   **2:** Gas exchange.   **3:** Surfactant.   **4:** Reduce the surface tension.
7.   **1:** Provides a patent airway; serves as a switching mechanism to route food into the posterior esophagus; voice
       production (contains vocal cords).   **2:** Elastic.   **3:** Hyaline;   **4:** The epiglottis has to be flexible to be able to flap
       over the glottis during swallowing. The more rigid hyaline cartilages support the walls of the larynx.

Figure 13-2:   8.   **Figure 13-3:**

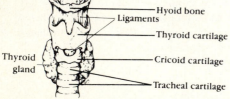

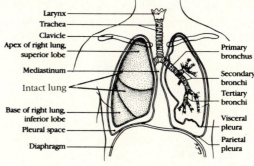

9.   **Figure 13-4:** The intact alveoli are the sac-like structures resembling grapes in part A; these should be colored yellow.
       The small vessels that appear to be spider webbing over their outer surface are the pulmonary capillaries. O_2 should
       be written inside the alveolar chamber and its arrow should move from the alveolus into the capillary. CO_2 should be
       written within the capillary and its arrow shown going from the capillary into the alveolar chamber.

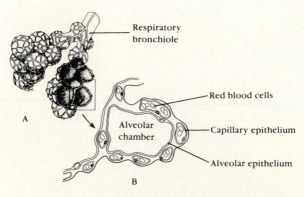

10. **1:** C or intrapleural pressure. **2:** A or atmospheric pressure. **3 and 4:** B or intrapulmonary pressure. **5:** C or intrapleural pressure. **6:** B or intrapulmonary pressure.

11. When the diaphragm is contracted, the internal volume of the thorax increases, the internal pressure in the thorax decreases, the size of the lungs increase, and the direction of air flow is into the lungs. When the diaphragm is relaxed, the internal volume of the thorax decreases, the internal pressure in the thorax increases, the size of the lungs decrease, and the direction of air flow is out of the lungs.

12. **1:** C or inspiration. **2:** D or internal respiration. **3:** E or ventilation. **4:** A or external respiration.

13. **1:** Sternocleidomastoid; pectoralis major. **2:** Transversus abdominis and external and internal obliques. **3:** Internal intercostals and latissimus dorsi.

14. **1:** Hiccup. **2:** Cough. **3:** Sneeze. **4:** Yawn.

15. **1:** E or tidal volume. **2:** A or dead space volume. **3:** F or vital capacity. **4:** D or residual volume. **5:** B or expiratory reserve volume.

16. **Figure 13-5:**

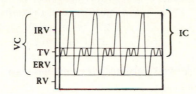

17. **1:** F. **2:** G. **3:** H. **4:** B. **5:** E. **6:** J. **7:** D. **8:** C. **9:** I.

18. **1:** Hemoglobin. **2:** Bicarbonate ion. **3:** Plasma. **4:** Oxygen.

19. **1:** Acidosis. **2:** ↑pH. **3:** Hyperventilation. **4:** ↑Oxygen. **5:** ↑CO_2 in blood. **6:** ↑PCO_2.

20. **1:** A or apneustic center, E or pneumotaxic center. **2:** C or expiratory center, D or inspiratory center. **3:** F or stretch receptors in lungs. **4:** B or chemoreceptors.

21. **1:** A or apnea. **2:** F or eupnea. **3:** D or dyspnea. **4:** G or hypoxia. **5:** E or emphysema. **6:** C or chronic bronchitis. **7:** B or asthma. **8:** C or chronic bronchitis, E or emphysema. **9:** H or lung cancer.

22. **1:** Infant respiratory distress syndrome (or hyaline membrane disease). **2:** Surfactant. **3:** Lower the surface tension of the watery film in the alveolar sacs. **4:** It keeps the lungs inflated so that gas exchange can continue.

23. **1:** 40. **2:** 12-18. **3:** Asthma. **4:** Chronic bronchitis. **5:** Emphysema or TB. **6:** Elasticity. **7:** Vital capacity. **8:** Infections, particularly pneumonia.

24. **1:** Nasal conchae. **2:** Pharyngeal tonsils. **3:** Nasopharynx. **4:** Mucus. **5:** Vocal fold. **6:** Larynx. **7:** Digestive. **8:** Epiglottis. **9:** Trachea. **10:** Cilia. **11:** Oral cavity. **12:** Primary bronchi. **13:** Left. **14:** Bronchiole. **15:** Alveolus. **16:** Red blood cells. **17:** Red. **18:** Oxygen. **19:** Carbon dioxide. **20:** Cough.

Chapter 14

1. **1:** Oral cavity. **2:** Digestion. **3:** Blood. **4:** Eliminated. **5:** Feces. **6:** Alimentary canal. **7:** Accessory.

2. **Figure 14-1:** The ascending, transverse, descending, and sigmoid colon are all part of the large intestine. The parotid, sublingual, and submandibular glands are salivary glands.

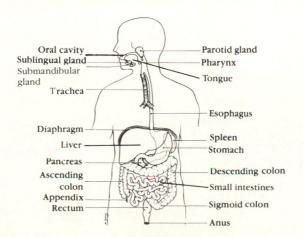

3. **Figure 14-2:** The frenulum should be red; the soft palate blue; the tonsils yellow; and the tongue pink.

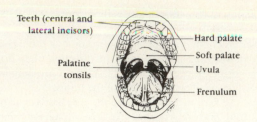

Teeth (central and lateral incisors)
Hard palate
Soft palate
Palatine tonsils
Uvula
Frenulum

4. **1:** A or Brunner's glands.   **2:** E or salivary glands.   **3:** D or pancreas.   **4:** C or liver.   **5:** B or gastric glands.

5. **1:** J or mesentery.   **2:** X or villi.   **3:** N or Peyer's patches.   **4:** P or plicae circulares.   **5:** L or oral cavity, U or stomach.   **6:** V or tongue.   **7:** O or pharynx.   **8:** E or greater omentum, I or lesser omentum, J or mesentery.   **9:** D or esophagus.   **10:** R or rugae.   **11:** G or haustra.   **12:** K or microvilli.   **13:** H or ileocecal valve.   **14:** S or small intestine.   **15:** C or colon.   **16:** W or vestibule.   **17:** B or appendix.   **18:** U or stomach.   **19:** I or lesser omentum.   **20:** S or small intestine.   **21:** Q or pyloric valve.   **22:** T or soft palate.   **23:** S or small intestine.   **24:** M or parietal peritoneum.   **25:** C or colon.   **26:** A or anus.   **27:** F or hard palate.   **28:** E or greater omentum.   **29:** D or esophagus.

6. **Figure 14-3:** On part B, the parietal cells should be colored red and the chief cells colored blue.

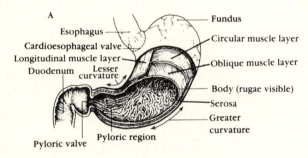

A
Esophagus
Cardioesophageal valve
Longitudinal muscle layer
Duodenum   Lesser curvature
Fundus
Circular muscle layer
Oblique muscle layer
Body (rugae visible)
Serosa
Greater curvature
Pyloric valve   Pyloric region

7. **Figure 14-4.**
    1: Mucosa.
    2: Muscularis externa.
    3: Submucosa.
    4: Serosa.

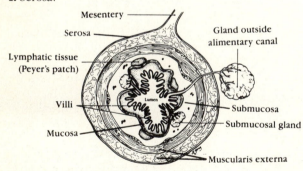

Mesentery
Serosa
Lymphatic tissue (Peyer's patch)
Villi
Mucosa
Gland outside alimentary canal
Lumen
Submucosa
Submucosal gland
Muscularis externa

8. **Figure 14-5:**

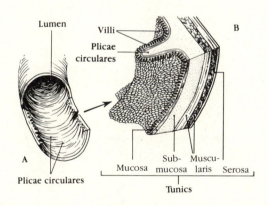

Lumen
Villi
Plicae circulares
B
A
Plicae circulares
Sub-mucosa
Mucosa
Muscu-laris
Serosa
Tunics

9. **Figure 14-6:**

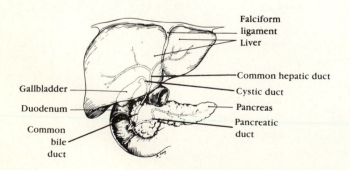

Falciform ligament
Liver
Gallbladder
Duodenum
Common bile duct
Common hepatic duct
Cystic duct
Pancreas
Pancreatic duct

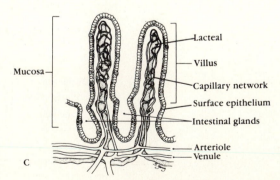

Lacteal
Villus
Capillary network
Surface epithelium
Intestinal glands
Mucosa
Arteriole
Venule
C

10.   **1:** Deciduous.   **2:** 6 months.   **3:** 6 years.   **4:** Permanent.   **5:** 32.   **6:** 20.   **7:** Incisors.   **8:** Canine.
     9: Premolars.   **10:** Molars.   **11:** Wisdom.

11.   **Figure 14-7.**   **1:** A or cementum.   **2:** C or enamel.   **3:** D or periodontal membrane.   **4:** B or dentin.   **5:** E or pulp.

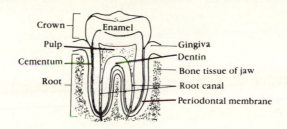

12.   **1:** Esophagus.   **2:** Rugae.   **3:** Gallbladder.   **4:** Cecum.   **5:** Circular folds.   **6:** Frenulum.   **7:** Palatine.   **8:** Saliva.
     9: Protein absorption.

13.   **1:** O or salivary amylase.   **2:** C or chewing.   **3:** M or psychologic stimulus.   **4:** I or mechanical stimulus.
     5: L or pepsin.   **6:** F or HCl.   **7:** K or mucus.   **8:** N or rennin.   **9:** D or churning.   **10:** E or disaccharases.
     11: A or bicarbonate-rich fluid.   **12:** H or lipases.   **13:** B or bile.

14.   **1:** G or peritonitis.   **2:** E or heartburn.   **3:** F or jaundice.   **4:** H or ulcer.   **5:** C or diarrhea.   **6:** D or gallstones.
     7: B or constipation.

15.   **1:** A or cholecystokinin, C or secretin.   **2:** B or gastrin.   **3:** A or cholecystokinin.   **4:** C or secretin.

16.   **1:** B or bread/pasta, D or cellulose, F or fructose, G or galactose, H or glucose, I or lactose, J or maltose, L or starch,
     M or sucrose.   **2:** F or fructose, G or galactose, H or glucose.   **3:** H or glucose.   **4:** I or lactose, J or maltose, M or
     sucrose.   **5:** L or starch.   **6:** D or cellulose.   **7:** C or cheese/cream, K or meat/fish.   **8:** A or amino acids.
     9: C or cheese/cream.   **10:** E or fatty acids.

17.   **1:** P.   **2:** A.   **3:** A.   **4:** P.   **5:** A.   Circle fatty acids.

18.   **1:** Deglutition.   **2:** Buccal or voluntary.   **3:** Involuntary.   **4:** Tongue.   **5:** Uvula.   **6:** Larynx.   **7:** Epiglottis.
     8: Peristalsis.   **9:** Cardioesophageal.   **10:** Peristalsis.   **11:** Segmental.   **12:** Segmental.   **13:** Mass peristalsis.
     14: Rectum.   **15:** Defecation.   **16:** Emetic.   **17:** Vomiting.

19.   **1:** B or carbohydrates.   **2:** C or fats.   **3:** A or amino acids.   **4:** C or fats.   **5:** C or fats.   **6:** A or amino acids.

20.   **1:** Chemical.   **2:** Catabolic.   **3:** Anabolic.   **4:** Albumin.   **5:** Clotting proteins.   **6:** Cholesterol.   **7:** Hyperglycemia.
     8: Glycogen.   **9:** Hypoglycemia.   **10:** Glycogenolysis.   **11:** Gluconeogenesis.   **12:** Detoxification.
     13: Phagocytic.

21.   **1:** TMR.   **2:** ↓Metabolic rate.   **3:** Child.   **4:** Fats.

22.   **1:** D or heat.   **2:** B or constriction of skin blood vessels, K or shivering.   **3:** A or blood.   **4:** F or hypothalamus.
     5: J or pyrogens.   **6:** C or frostbite.   **7:** I or radiation, H or perspiration.   **8:** G or hypothermia.
     9: E or hyperthermia.

23.   **1:** Glucose.   **2:** Oxygen.   **3:** Water.   **4:** Carbon dioxide.   **5:** ATP.   **6:** Monosaccharides.   **7 and 8:** (in any order):
     Acetoacetic acid; Acetone.   **9:** Ketosis.   **10:** Essential.   **11:** Ammonia.   **12:** Urea.

24.   **1:** B or alimentary canal.   **2:** A or accessory organs.   **3:** D or cleft palate/lip.   **4:** N or tracheoesophageal fistula.
     5: E or cystic fibrosis.   **6:** H or PKU.   **7:** K or rooting.   **8:** M or stomach.   **9:** C or appendicitis.   **10:** G or gastritis,
     O or ulcers.   **11:** I or periodontal disease.   **12:** J or peristalsis.

25.   **1:** Mucosa.   **2:** Vestibule.   **3:** Tongue.   **4:** Salivary amylase.   **5:** Peristalsis.   **6:** Esophagus.   **7:** Larynx.
     8: Epiglottis.   **9:** Stomach.   **10:** Mucus.   **11:** Pepsin.   **12:** Hydrochloric acid.   **13:** Pyloric.   **14:** Lipase.
     15: Pancreas.   **16:** Villi.   **17:** Ileocecal.

Chapter 15

1.   **1:** Nitrogenous.   **2:** Water.   **3:** Acid-base.   **4:** Kidneys.   **5:** Ureters.   **6:** Peristalsis.   **7:** Bladder.   **8:** Urethra.   **9:** 8.   **10:** $1\frac{1}{2}$

2.   **Figure 15-1:**

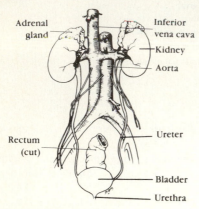

3.   **Figure 15-2:** The fibrous membrane surrounding the kidney is the *renal capsule*; the basinlike *pelvis* is continuous with the ureter; a *calyx* is an extension of the pelvis; *renal columns* are extensions of cortical tissue into the medulla. The cortex contains the bulk of the nephron structures; the striped-appearing medullary pyramids are primarily formed by collecting ducts.

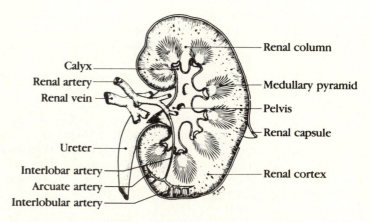

4.   **1:** Kidney.   **2:** Forms urine.   **3:** Continuous with renal pelvis.   **4:** Glomerulus.   **5:** Collecting tubule.   **6:** Cortical nephrons.   **7:** Collecting duct.   **8:** Glomeruli.

5.   **Figure 15-3.**   **1:** Bowman's capsule.   **2:** Afferent arteriole.   **3:** Efferent arteriole.   **4:** Interlobular artery.   **5:** Interlobular vein.   **6:** Arcuate artery.   **7:** Arcuate vein.   **8:** Interlobar artery.   **9:** Interlobar vein.   **10:** Loop of Henle.   **11:** Collecting tubule.   **12:** DCT.   **13:** PCT.   **14:** Peritubular capillaries.   **15:** Glomerulus. Relative to the coloring instructions, #1 is green, #15 is red, #14 is blue, #11 is yellow, and #13 is orange.

6.   **1:** B or urethra.   **2:** A or bladder.   **3:** A or bladder.   **4:** B or urethra.   **5:** B or urethra, C or ureter.   **6:** B or urethra.   **7:** C or ureter.   **8:** A or bladder, C or ureter.   **9:** B or urethra.   **10:** B or urethra.

7.   **1:** A or cystitis.   **2:** C or hydronephrosis.   **3:** F or uremia.   **4:** E or pyelonephritis.   **5:** B or diabetes insipidus.   **6:** D or ptosis.

8.   **1:** Afferent.   **2:** Efferent.   **3:** Plasma.   **4:** Diffusion.   **5:** Active transport.   **6:** Microvilli.   **7:** Secretion.   **8:** Diet.   **9:** Cellular metabolism.   **10:** Urine output.   **11:** 1-1.8.   **12:** Urochrome.   **13:** Urea.   **14:** Uric acid.   **15:** Creatinine.   **16:** Lungs.   **17:** Perspiration.   **18:** Decreases.   **19:** Dialysis (artificial kidney).

9.   **1:** Aldosterone.   **2:** Secretion.   **3:** $\uparrow K^+$ reabsorption.   **4:** $\downarrow BP$.   **5:** $\downarrow K^+$ retention.   **6:** $\uparrow HCO_3^-$ in urine.

10.   **1:** A.   **2:** B.   **3:** A.   **4:** A.   **5:** B.

11.   **1:** D.   **2:** D.   **3:** I.   **4:** D.   **5:** I.   **6:** I.

12.   **1:** B.   **2:** A.   **3:** A.   **4:** C.   **5:** B.   **6:** C.   **7:** A.   **8:** A.   **9:** A.   **10:** C.   **11:** A.   **12:** B.

13.   **1:** Hematuria; bleeding in urinary tract.   **2:** Ketonuria; diabetes mellitus, starvation.   **3:** Albuminuria; glomerulonephritis, pregnancy.   **4:** Pyuria; urinary tract infection.   **5:** Bilirubinuria; liver disease.   **6:** (No official terminology); kidney stones.   **7:** Glycosuria; diabetes mellitus.

14.   **1:** All reabsorbed by tubule cells.   **2:** Usually does not pass through the glomerular filter.

15.   **1:** Micturition.   **2:** Stretch receptors.   **3:** Contract.   **4:** Internal urethral sphincter.   **5:** External urethral sphincter.   **6:** Voluntarily.   **7:** Approximately 600.   **8:** Incontinence.   **9:** Infants, toddlers.   **10:** Emotional/neural problems.   **11:** Pressure (pregnancy).   **12:** Urinary retention.   **13:** Prostate.

16.   Check 1, 3, 4.

17. **Figure 15-4:**
 Black arrows: Site of filtrate formation is the glomerulus. Arrows leave the glomerulus and enter Bowman's (glomerular) capsule. **Red arrows:** Major site of amino acid and glucose reabsorption. Shown going *from* the PCT interior and passing through the PCT walls to the capillary bed surrounding the PCT. Nutrients *leave* the filtrate. **Green arrows:** At side of ADH action. Arrows (indicating water movement) shown *leaving* the interior of the collecting tubule and passing through the walls to enter the capillary bed surrounding that tubule. Water *leaves* the filtrate. **Yellow arrows:** Site of aldosterone action. Arrows (indicating Na⁺ movement) *leaving* the ascending limb of the loop of Henle, the DCT, and the collecting tubule and passing through their walls into the surrounding capillary bed. Na⁺ *leaves* the filtrate. **Blue arrows:** Site of tubular secretion. arrows shown *entering* the PCT to enter the filtrate.

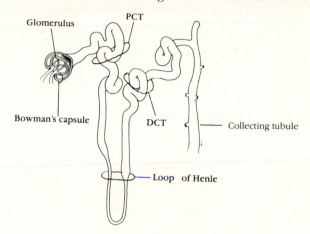

18. **1:** Male adult. **2:** Lean adult. **3:** ICF. **4:** Nonelectrolyte. **5:** $\downarrow$H⁺ secretion. **6:** $\uparrow$ADH. **7:** $\uparrow$K⁺ reabsorption. **8:** $\downarrow$BP.
19. **1:** Placenta. **2:** Polycystic disease. **3:** Hypospadias. **4:** Males. **5:** Bladder. **6:** 18-24. **7:** Glomerulonephritis. **8:** Antigen-antibody. **9:** Protein. **10:** Blood. **11:** Arteriosclerosis. **12:** Tubule. **13:** Urgency. **14:** Frequency.
20. **1:** Tubule. **2:** Renal. **3:** Afferent. **4:** Glomerulus. **5:** Bowman's capsule. **6:** Plasma. **7:** Proteins. **8:** Loop of Henle. **9:** Microvilli. **10:** Reabsorption. **11:** Glucose. **12:** Amino acids. **13:** 7.4 (7.35-7.45). **14:** Nitrogenous. **15:** Sodium. **16:** Potassium. **17:** Urochrome. **18:** Antidiuretic hormone. **19:** Collecting duct. **20:** Pelvis. **21:** Peristalsis. **22:** Urine. **23:** Micturition. **24:** Urethra.

Chapter 16

1. Seminiferous tubule→Rete testis→Epididymis→Ductus deferens.
2. Deepening voice; formation of a beard and increased hair growth all over body particularly in axillary/genital regions; enlargement of skeletal muscles; increased density of skeleton.
3. **1:** E or penis. **2:** K or testes. **3:** C or ductus deferens. **4:** L or urethra. **5:** A or Cowper's glands, G or prostate gland, H or seminal vesicles, K or testes. **6:** I or scrotum. **7:** B or epididymis. **8:** F or prepuce. **9:** G or prostate gland. **10:** H or seminal vesicles. **11:** A or Cowper's glands. **12:** J or spermatic cord.
4. **Figure 16-1:** The spongy tissue is the penis; the duct system that serves the urinary system also is the urethra; the structure providing ideal temperature conditions is the scrotum; the prepuce is removed at circumcision; the glands producing a secretion that contains sugar are the seminal vesicles; the ductus deferens is cut or cauterized during vasectomy.

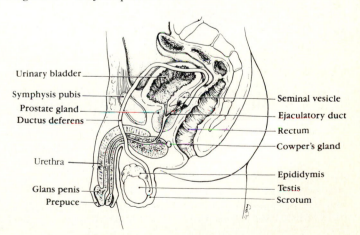

5. **Figure 16-2:** The site of spermatogenesis is the seminiferous tubule. Sperm mature in the epididymis. The fibrous coat is the tunica albuginea.

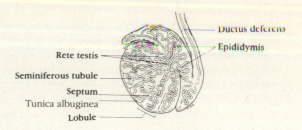

Ductus deferens
Epididymis
Rete testis
Seminiferous tubule
Septum
Tunica albuginea
Lobule

6. **1:** Spermatogonium.   **2:** Secondary spermatocyte, Sperm, Spermatid.   **3:** Secondary spermatocyte.   **4:** Spermatid.
   5: Sperm.   **6:** FSH, Testosterone.

Figure 16-3:

7. **Figure 16-4:**

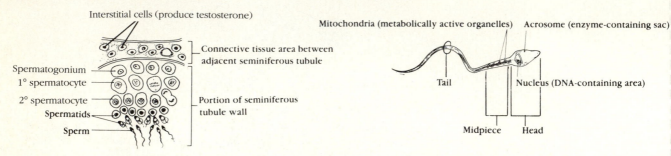

Interstitial cells (produce testosterone)

Spermatogonium
1° spermatocyte
2° spermatocyte
Spermatids
Sperm

Connective tissue area between adjacent seminiferous tubule

Portion of seminiferous tubule wall

Mitochondria (metabolically active organelles)   Acrosome (enzyme-containing sac)

Tail

Nucleus (DNA-containing area)

Midpiece   Head

8. **1:** A or mitosis.   **2:** B or meiosis.   **3:** C or both mitosis and meiosis.   **4:** A or mitosis.   **5:** B or meiosis.
   6: A or mitosis.   **7:** A or mitosis.   **8:** B or meiosis.   **9:** C or both mitosis and meiosis.   **10:** B or meiosis.
   11: B or meiosis.

9. **1:** Uterus.   **2:** Vagina.   **3:** Uterine, or fallopian tube.   **4:** Clitoris.   **5:** Uterine tube.   **6:** Hymen.   **7:** Ovary.
   8: Fimbriae.

10. **1:** B or primary oocyte.   **2:** C or secondary oocyte.   **3:** C or secondary oocyte.   **4:** D or ovum.

11. **Figure 16-5:** The egg travels along the fallopian tube after it is released from the ovary. The round ligament helps to anchor the uterus. The ovary forms hormones and gametes, and the labia majora are the homologue of the male scrotum.

12. **Figure 16-6:** The clitoris should be colored blue; the hymen yellow; and the vaginal opening red.

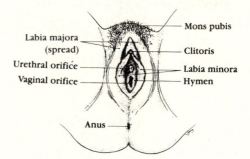

Labia majora (spread)
Urethral orifice
Vaginal orifice
Mons pubis
Clitoris
Labia minora
Hymen
Anus

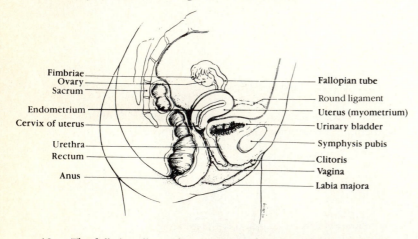

Fimbriae
Ovary
Sacrum
Endometrium
Cervix of uterus
Urethra
Rectum
Anus

Fallopian tube
Round ligament
Uterus (myometrium)
Urinary bladder
Symphysis pubis
Clitoris
Vagina
Labia majora

13. The follicle cells produce estrogen; the corpus luteum produces progesterone; and oocytes are the central cells in all follicles. Event A = ovulation.   **1:** No.   **2:** Peritoneal cavity.
    3: When sperm penetration occurs.   **4:** Ruptured (ovulated) follicle.   **5:** One ovum; three polar bodies.   **6:** Males produce four spermatids → four sperm.   **7:** They deteriorate.   **8:** They lack nutrient-containing cytoplasm.   **9:** Menopause.

Figure 16-7:

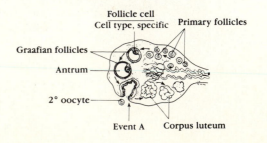

Follicle cell
Cell type, specific
Primary follicles
Graafian follicles
Antrum
2° oocyte
Event A
Corpus luteum

14. Because of this structural condition, many "eggs" (oocytes) are lost in the peritoneal cavity; therefore, they are unavailable for fertilization. The discontinuity also provides infectious microorganisms with access to the peritoneal cavity, possibly leading to PID.

15. **1:** FSH (follicle-stimulating hormone).   **2:** LH (luteinizing hormone).   **3:** Estrogen and progesterone.   **4:** Estrogen. **5:** LH.   **6:** LH.

16. Appearance of axillary/pubic hair, development of breasts, widening of pelvis, onset of menses.

17. **1:** A or estrogen, B or progesterone.   **2:** B or progesterone.   **3:** A or estrogen.   **4:** B or progesterone. **5 and 6:** A or estrogen.

18. **Figure 16-8:** From left to right on part C the structures are: the primary follicle, the growing follicle, the graafian follicle, the ovulating follicle, and the corpus luteum. In part D, menses is from day 0 to day 4, the proliferative stage is from day 4 to day 14, and the secretory stage is from day 14 to day 28.

19. **Figure 16-9:** The alveolar glands should be colored blue, and the rest of the internal breast, excluding the duct system, should be colored yellow.

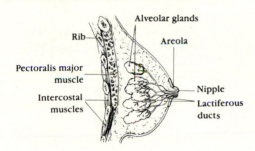

20. **Figure 16-10, 1:** Fertilization (sperm penetration).   **2:** 2° oocyte.   **3:** Cleavage.   **4:** Blastocyst (chorionic vesicle). **5:** Gastrulation.

21. **1:** F or inner cell mass.   **2:** G or placenta.   **3:** B or chorionic villi, C or endometrium.   **4:** A or amnion.   **5:** H or umbilical cord.   **6:** B or chorionic villi.   **7:** E or fetus.   **8:** G or placenta.   **9:** D or fertilization.   **10:** I or zygote.

22. The blastocyst and then the placenta release HCG, which is like LH and sustains the function of the corpus luteum temporarily.

23. Oxytocin and prostaglandins.

24. **1:** Prolactin.   **2:** Oxytocin.

25. Check 1, 3, 5, 9, 10, 11, 12, 13.

26. False labor (irregular, ineffective uterine contractions). This occurs because rising estrogen levels make the uterus more responsive to oxytocin and antagonize progesterone's quieting influence on the myometrium.

27. **1:** Dilation stage: The period from the beginning of labor until full dilation (approx. 10 cm diameter) of the cervix; the longest phase.   **2:** Expulsion stage: The period from full dilation to the birth (delivery).   **3:** Placental stage: Delivery of the placenta, which follows delivery of the infant.

28. **Figure 16-11:**

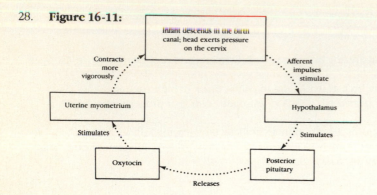

29. Each pass forces the baby farther into the birth passage. The cycle ends with the birth of the baby.

30. The response to the stimulus enhances the stimulus. For example, the more a baby descends into the pelvis and stretches the uterus, the more oxytocin is produced and the stronger the contractions become.

31. **1:** Y and X.   **2:** 2 Xs.   **3:** Male external genitalia and accessory structures.   **4:** Female external genitalia and duct system.   **5:** Cryptorchidism.   **6:** *Escherichia coli*.   **7:** STDs or venereal disease.   **8:** Yeast infections.   **9:** PID (pelvic inflammatory disease).   **10:** Venereal disease microorganisms.   **11:** Breast.   **12:** Cervix of the uterus.   **13:** Pap smear.   **14:** Menopause.   **15:** Hot flashes.   **16:** Declines.   **17:** Rise.   **18:** Estrogen.   **19:** Vaginal.   **20:** Prostate gland.   **21:** Urinary.   **22:** Reproductive.

32. **1:** Uterus.   **2:** Ovary.   **3:** Fimbriae.   **4:** Ovulation.   **5:** 2° oocyte.   **6:** Follicle.   **7:** Peristalsis.   **8:** Cilia.   **9:** Sperm.   **10:** Acrosomes.   **11:** Meiotic.   **12:** Ovum.   **13:** Polar body.   **14:** Dead.   **15:** Fertilization.   **16:** Zygote (fertilized egg).   **17:** Cleavage.   **18:** Endometrium.   **19:** Implantation.   **20:** Vagina.

Notes